Methods of Mine Timbering

California State Mining Bureau

*Compiled by the staff of
the California State Mining Bureau*

with an introduction by Kerby Jackson

*This work contains material that was originally published in 1896 by
the US Department of Interior.*

*This publication was created and published for the public benefit,
utilizing public funding and is within the Public Domain.*

*This edition is reprinted for educational purposes
and in accordance with all applicable Federal Laws.*

Introduction Copyright 2014 by Kerby Jackson

Introduction

It has been over a century since the California State Mining Bureau released it's important publication "Methods of Mine Timbering". First released in 1896, this work has been unavailable to the mining community since those days, with the exception of expensive original collector's copies and poorly produced digital editions.

It has often been said that *"gold is where you find it"*, but even beginning prospectors understand that their chances for finding something of value in the earth or in the streams of the Golden West are dramatically increased by going back to those places where gold and other minerals were once mined by our forerunners. Despite this, much of the contemporary information on local mining history that is currently available is mostly a result of mere local folklore and persistent rumors of major strikes, the details and facts of which, have long been distorted. Long gone are the old timers and with them, the days of first hand knowledge of the mines of the area and how they operated. Also long gone are most of their notes, their assay reports, their mine maps and personal scrapbooks, along with most of the surveys and reports that were performed for them by private and government geologists. Even published books such as this one are often retired to the local landfill or backyard burn pile by the descendents of those old timers and disappear at an alarming rate. Despite the fact that we live in the so-called "Information Age" where information is supposedly only the push of a button on a keyboard away, true insight into mining properties remains illusive and hard to come by, even to those of us who seek out this sort of information as if our lives depend upon it. Without this type of information readily available to the average independent miner, there is little hope that our metal mining industry will ever recover.

This important volume and others like it, are being presented in their entirety again, in the hope that the average prospector will no longer stumble through the overgrown hills and the tailing strewn creeks without being well informed enough to have a chance to succeed at his ventures.

Kerby Jackson
Josephine County, Oregon
October 2014

METHODS OF MINE-TIMBERING.

By W. H. Storms, M.E., Assistant in the Field.

The excavation of any considerable amount of earth or rock beneath the surface of the ground usually necessitates that the roof, and not infrequently the sides, of such excavation be sustained artificially to prevent caving. In these later years the size of underground excavations is so great, as compared with those formerly made, that the ingenuity and skill of the miner are taxed to the utmost limit. So successful, however, have miners become in devising novel methods to meet daily exigencies that the obstacles usually encountered in mining—among which are flaky rock roofs; soft, running ground; floods of water, sometimes scalding hot; and, worst of all, swelling ground, with heavy pressure from all sides, including the bottom—have mostly been successfully overcome.

Before the "square-set" system came into use, the ingenious placing of posts, caps, and "stulls" constituted the only method of timbering, the multitude of conditions met in the mines making the combinations almost endless. Where veins occur in firm, solid rock, being perpendicular, or near it, the danger of caving is greatly lessened, and the amount of timber required is reduced to a minimum; but these conditions are exceptional.

Veins dip at all angles between perpendicular and horizontal, and vary greatly in width. Moreover, the character of the wall rocks, as well as of the ore itself, is so variable that unexpected problems are encountered daily. The shape of large bodies of ore is a matter of great importance in determining the system of timbering to be employed.

Through all this variety and change in form, dip, size, and character of the vein, or deposit, and in the inclosing walls, certain established principles are followed in sustaining the roof and sides, the constant aim being to prevent caving; and to avoid such a catastrophe timbers are placed with a view of holding the rock-masses in place, and always in such a manner as to receive the strain directly. Those timbers which reach from wall to wall of an inclined vein (stulls) are not set at a right angle to the pitch of the hanging wall, but at a somewhat higher angle. The reason is obvious, for if placed at a right angle, should a subsidence of the wall occur, the timber, partaking of this movement, at its upper end, would then have a tendency to fall of its own weight; whereas, if set originally above an angle of 90° with the hanging wall, the subsidence of that wall only serves to wedge and hold the timber support more firmly, when it must bend and break before falling. When properly placed, stull timbers usually give sufficient warning of their weakness to permit of placing additional timbers and subsequently the removal and replacement of old or weak timbers.

To make an understanding of the various practices more comprehensive a number of drawings have been introduced in this article. As a matter of course, certain exigencies are likely to arise, the character of

6   METHODS OF MINE-TIMBERING.

which has not been anticipated, but in all cases the principles applied remain the same, and it is thought that the conditions most likely to occur have been treated fully enough to meet the demands of metalliferous mining generally.

KINDS OF TIMBER USED.

In some instances solid masonry is built to sustain certain portions of mines, and in late years iron has been introduced as a substitute for timber, but in American mines timber is most commonly employed. Ordinarily the location of the mine determines the kind of timber. Pine of various kinds is more extensively used than any other. Sugar

Fig. 1.

pine and spruce are preferred when obtainable, but yellow pine is by far the most common. Oak is gladly taken when good sticks of sufficient size can be found. There are oak timbers in mines in Mariposa County, apparently sound, that have been in place for twenty years. They certainly outlast any other timber. Cottonwood is sometimes used when no other is obtainable, but is not at all desirable. In the desert in this State, in parts of Arizona and Nevada, miners take any timber they can get, even resorting to the yucca, which answers quite well for a time in that dry climate, when the pressure is not too great. At Silver Reef, in San Bernardino County, green yuccas were used in timbering a drift; they still stood in fair condition, after having been in place five years. The Superintendent of the Gover Mine, Amador County, California, has commenced some experiments with spruce and sugar pine, placing them side by side in the same set in the lower levels of his mine, to test their durability.

TUNNELS AND DRIFTS.

The methods employed to sustain the roof and sides of tunnels and drifts are numerous, the existing conditions determining the method. Often the rock is sufficiently firm to stand without timber, but at times the conditions are annoying and dangerous.

Where the pressure is entirely overhead an upright post is set on either side of the tunnel, usually spread somewhat at the bottom, but otherwise always at right angles to the roof. On these posts is placed a cross-piece, called a cap. Drifts are frequently run on an incline, particularly in blanket veins, and also in fissures having a low angle of dip. When it may occur that the floor of a tunnel, drift, or cross-cut does not afford a firm foundation for the posts, as in soft, wet fissures, when not in ore, or as is often the case at the entrance to a mine, a cross-piece or sill is first laid and the posts set upon it. A second cross-piece (the

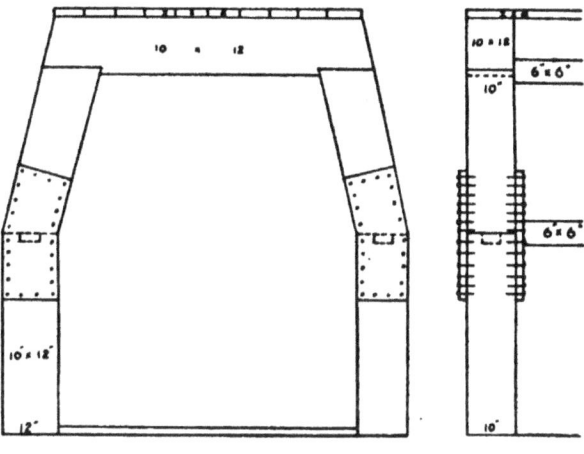

Fig. 2.

cap) is then inserted at the top, the ends resting upon the posts, the cap being employed in all cases whether the sill is necessary or not.

In working ground that is fairly firm, particularly in the drift gravel mines of California, a system of posts and "breasting caps" is used. (See Fig. 12.) This consists of a piece of timber, hewn or split, $2\frac{1}{2}'$ long, 1' wide, and 3" or 4" in thickness (the cap), which is placed against the roof, and a post of the necessary length is set beneath it, being driven into a perpendicular position by blows of a heavy hammer or maul. It is a cheap and secure method of timbering small drifts, and is often employed in large ones, the breast extending entirely across the channel. Lagging, placed at right angles to the cap, may be driven in above it when necessary.

METHODS OF MINE-TIMBERING.

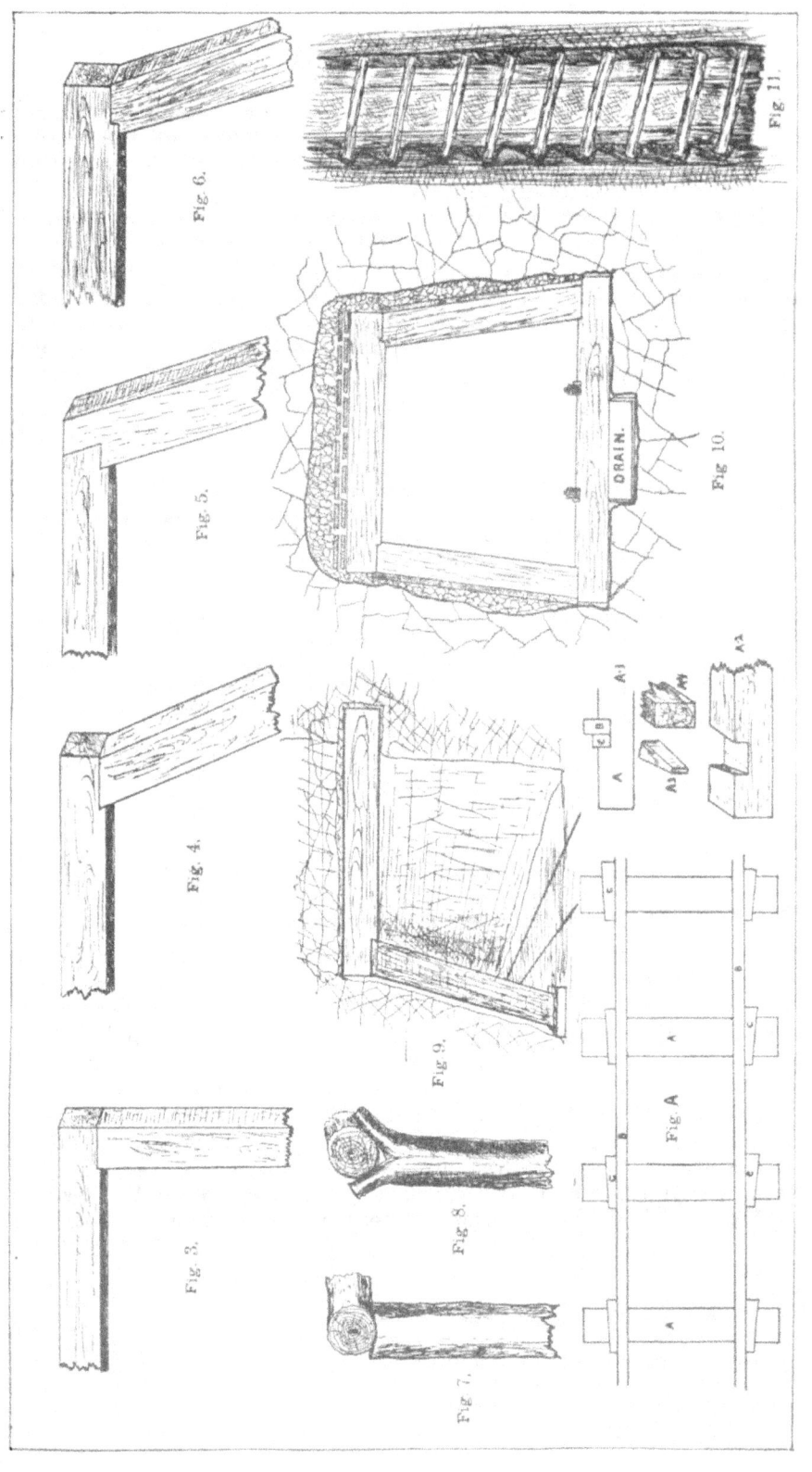

METHODS OF MINE-TIMBERING.

Very often pressure is exerted from the sides as well as from the top of a drift. In such cases the timbers are framed with a view to binding them more firmly together when in place. (See Figs. 10 and 19.)

Figs. 3, 4, 5, and 6 show several styles of framing timbers for drifts. Of these, Fig. 6 is undoubtedly the best. Better than any of these is the

Fig. 12.   Fig. 13.

beveled notch (Fig. 13), which greatly reduces the liability of the timbers splitting. It has come quite extensively into use of late years. When properly framed and set there is little danger of slipping. Still another, and no doubt the best method of all, has lately come into practice. It is that of nailing a 2" plank on the under side of the cap. By this

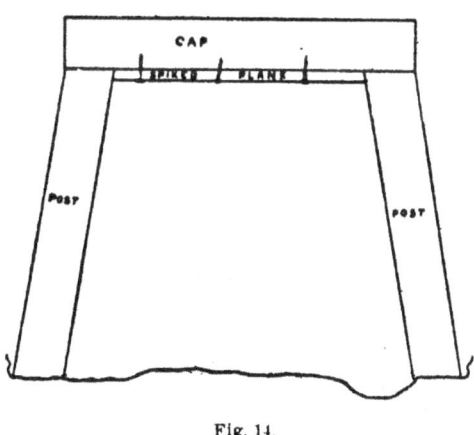

Fig. 14.

device the fullest strength of the timbers is obtained, with no probability of splitting.

When round timbers are used they should always be stripped of the bark, or they will speedily decay. The manner of framing timbers is the same, whether they be round, hewn, or sawed.

Figs. 7 and 8 show a method of placing posts and caps. It is seldom seen now. There is nothing about it to recommend it. Oak is the only timber that may be safely employed in this manner, pine being too soft and quite certain to split.

Where only one side and the roof of a drift need support, the post and cap (Fig. 9) is sometimes employed. One end of the cap rests upon the post, the other on a shelf, or niche, cut in the opposite wall.

LAGGING.

When the roof or side of a tunnel is loose and shows any tendency to cave, spawl off, or run, lagging must be employed. Lagging is the name given to strips of wood 4' to 6' long, 6" to 8" wide, and 2" to 2½" thick. They are usually split from pine logs, but sometimes 2" plank 6" in width is substituted. In large shafts in heavy ground 3" plank is sometimes employed. The methods of driving lagging are shown in Figs. 16 and 18. The pieces of lagging are inserted over the top of the

Fig. 15.

cap, the ends pointed upward a few degrees. They are driven forward as the work of excavation progresses, when there is danger of caving. Not infrequently ground will stand for many hours and sometimes for months before caving, but it is cheaper to timber very soon after the excavation has been made, in order to keep the ground in normal condition, giving it no chance to cave.

The two systems shown in Figs. 16 and 18, while the same in principle, differ materially in detail. In Fig. 16 the lagging is inserted between two caps, which are separated by wedge-shaped blocks, one of which is placed in the center and one at either end. (See between A and C, Fig. 17.) The lagging is driven forward as explained. If the ground is very heavy, a "false set" (Fig. 18) is set up and the ends of the lagging rest upon it. The excavation progressing and the lagging being driven well forward, the next set is put in position and the lagging driven "home"; that is, until the forward ends find a secure resting place on the true set. The false set is then knocked out, and the same procedure gone through with the next set.

The only difference between Fig. 16 and Fig. 18 is that in the former there are two caps, as explained above, while in Fig. 18 the lagging is

METHODS OF MINE-TIMBERING.

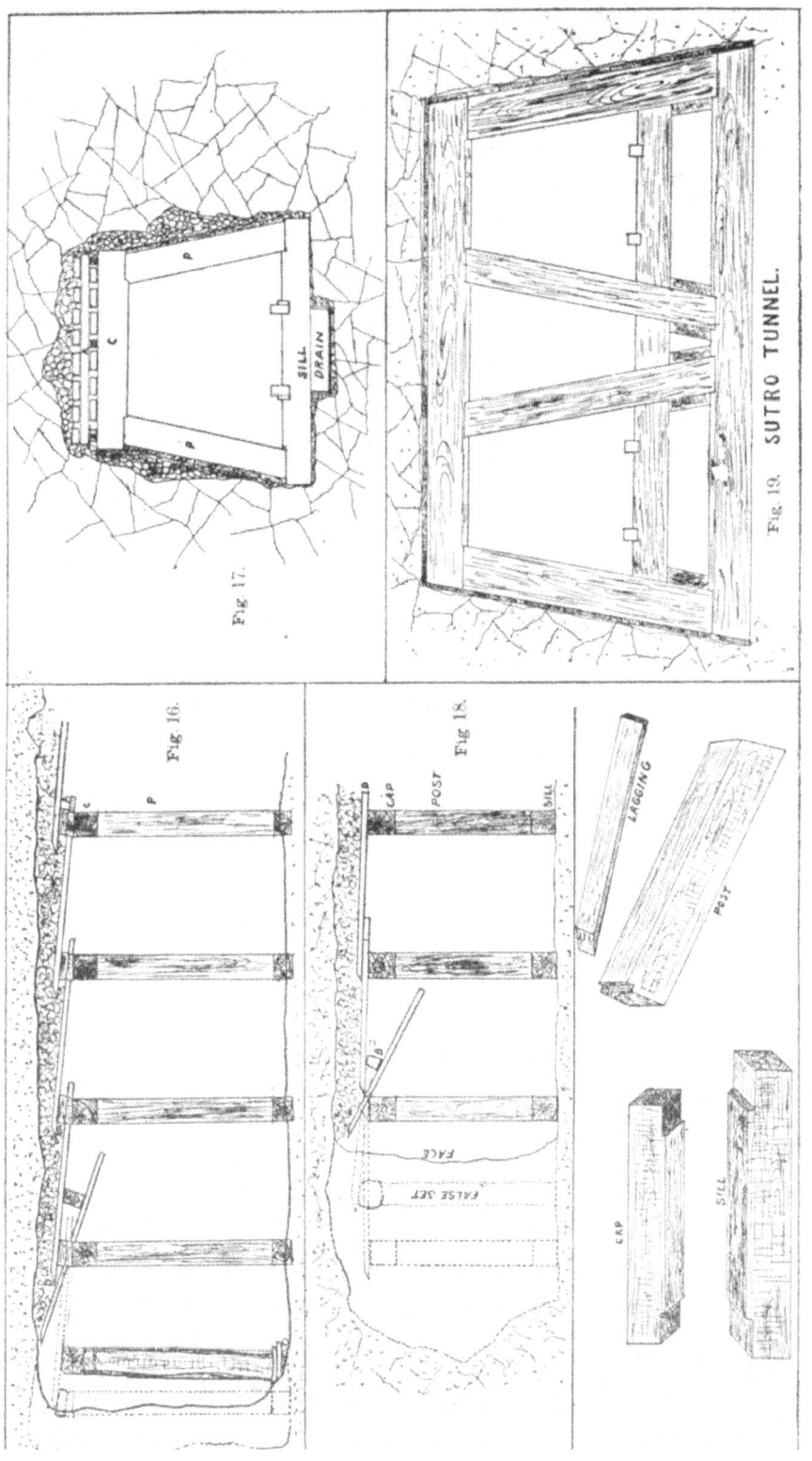

inserted beneath the forward ends of the next set back. In each case the lagging is kept pointed slightly upward by the insertion of a block of wood, shown at B, Fig. 18. When the forward ends of the lagging rest on the false set, this block may be allowed to drop out. The system shown in Fig. 16 can be employed much more advantageously and work progress more rapidly in very heavy ground, than when that shown in Fig. 18 is used, which does very well in lighter ground. Where the ground is very bad the lagging must be kept driven as far forward as possible. By observing care in this matter serious runs are sometimes prevented. Sills are only employed when the bottom of the tunnel does not afford a firm foundation for the posts.

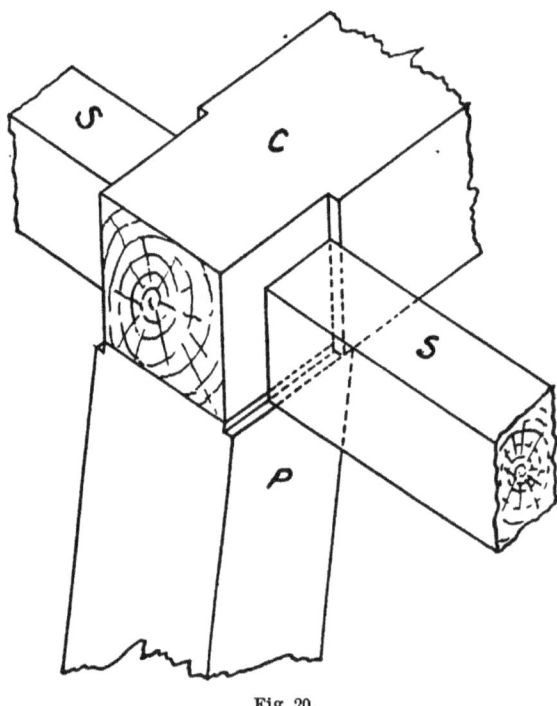

Fig. 20.

Sprags are sometimes employed in drift sets to prevent lateral movement of the timber. Fig. 20 shows a method of framing designed to prevent the sprag S from being forced out of position by side pressure, the end resting upon a shoulder on the post P, and against one on the cap C.

For lagging, spruce, yellow pine, and tamarack are much used, but the sugar pine of California has no superior for toughness and durability. Lagging should not be too strong, for in the event of extreme weight it should bend and give notice of impending danger. The miner may then relieve the pressure by cutting away a portion and reinforcing the timbers, thus saving the more expensive framed timbers and perhaps preventing a serious cave.

Caps and posts are all sizes, from 4"x 4" or 4"x 6" up to 24"x 30", according to the size of the excavation and the character of the ground. Caps should be free from knots and checks as far as possible. Less care is necessary in the selection of posts, though all timbers should be of good, sound wood. Sills extend somewhat beyond the posts which rest upon them. A shallow notch is usually cut in the sill to admit the post, the bottom of which is cut at right angles to its sides. (See Fig. 21.)

The greater care taken in framing and setting-up mine timbers the less danger there will be of collapse in the future. The tools necessary to secure this exactness are a plumb-bob and a steel square. A spirit-level is also very useful.

TRACK-LAYING.

To construct a track, cross-ties, made of 3"x 4" scantling, are often laid on the floor of the tunnel or drift, and to these are spiked "T" rails. When flat iron is preferred, a durable track for permanent use is made by setting long 2"x 4" scantling in slots sawed in the cross-ties, the long strips being secured by driving in wedges at the side. The details of track construction are shown in Figs. A, A1, A2, A3, and A4,

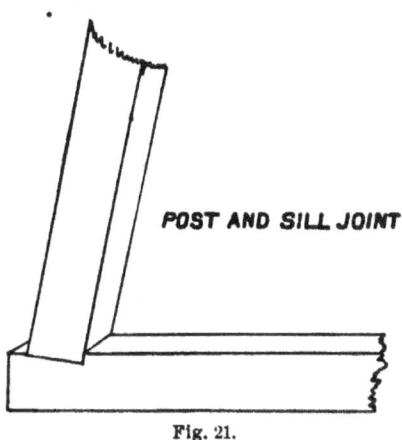

Fig. 21.

which are on the same plate as Figs. 3 to 11. Upon the scantling-stringers may be spiked the flat-iron, or if desired, "T" rail. The scantling should not be spiked to the cross-ties, as nails are quickly corroded by the mineral vapors and waters of mines.

DRAINAGE.

Provision for drainage should always be made at the very commencement of opening a mine, for, though the tunnel may be dry at its mouth, when it has been driven a long way into the mountain more or less water is nearly always encountered. A drain or trench should always be cut in the center or at the sides of every tunnel or drift. Illy drained workings cause the timbers to rot quickly, and also endanger the health of the miners. Neglect to provide drainage very often results in the necessity of retimbering, which expense might otherwise have been avoided for a long time.

Prospectors, in their haste to advance work as speedily and as cheaply as possible, frequently fail to timber properly and to provide for drainage, with disastrous results in the future. Many months of laborious toil are too often lost in this way, to say nothing of the loss of life and limb by the unfortunate miner caught in a cave which might have been easily avoided.

SWELLING GROUND.

One of the greatest difficulties with which miners have to contend is the swelling of the rock-masses into which their excavations have penetrated. Often the force or pressure against timbers caused by the swelling of the ground is irresistible. It is a common feature of many of the Mother Lode mines in California, particularly in Tuolumne, Calaveras, and Amador Counties. Swelling bedrock is quite common in the gravel drift mines of California. All Comstock miners know what swelling ground is. It is one of the most serious obstacles with which they have to contend.

In a general way it may be said that the only recourse is to timber in the most substantial manner, and then, by frequently, or as often as necessary, cutting out a portion of the heavy ground and relieving the pressure, the timbers may be kept in place and the excavation kept open. Fig. 22 represents a cross-section of a tunnel where this trouble in its worst form was encountered. By setting timbers in the manner shown in the cut, placing the sets close together, and relieving the ground by the removal of the encroaching portion from time to time, the trouble is reduced to a minimum. In the Hardenburg Mine, Amador County, swelling ground caused a great deal of trouble. The 900-foot level is run in a zone of crushed foliated black slate, which, on the foot-wall side of the drift, when first broken, appears firm and solid, but in a few days it commences to spawl off and to noticeably encroach upon the drift. It continues to swell, displacing timbers or breaking them, and causing no end of trouble. Now such places are timbered with 18" and 20" round timber, and somewhat loosely lagged. But few days pass before it is necessary to take out lagging and cut away the swelled ground.

In running the main tunnel of the Hidden Treasure Mine at Forest Hill, in Placer County, swelling bedrock was encountered. Mr. Ross E. Browne, E.M., in his article, "The Ancient River-Beds of the Forest Hill Divide" (*vide* X Report, State Mineralogist of California), says of this occurrence:

"The pressure of the gravel is not great, but the swelling bedrock has been a source of trouble, driving the legs of the timber-set inward and crushing the cap. After many unsuccessful attempts to overcome this difficulty, the legs were given an increasingly greater bottom spread, until finally it was found that they remained stationary. The swelling bedrock is removed from time to time and the track adjusted. The accompanying cut (Fig. 22) shows the form of tunnel timber-set now used in bad swelling ground. Sets are first put in 4' apart, and in the course of a few months center sets are placed between these. Timber-sets on this plan have now (1879) been in place three years, and are still in good condition. In 8,500 feet length of tunnel there are about 4,000 sets of timbers. Two men are kept constantly employed in easing and repairing the sets and adjusting the track."

Some of the drift mines on Sugar Loaf Mountain, near Nevada City, California, that were worked twenty-five years ago, were timbered in a very peculiar and unusual manner, owing to the swelling of the bedrock. Massive timbers had been placed time and again, only to be forced out of place or broken. At last the method here described was introduced, and found to answer every requirement most admirably. It was subsequently tried in some of the other mines in the neighborhood, with equally satisfactory results. Once firmly placed, the timbers were never again renewed, standing until the mine was worked out.

The plan adopted was as follows: A drift of the usual form was run and heavily timbered, being well lagged overhead. The sets were placed 5' from center to center. As the work of excavating the drift proceeded, a triangular section was cut out of each side of the drift between the posts of the two adjoining sets. These two posts formed the base of a triangle, the apex being directly opposite the center of the base. At the apex a post was set, the center of which was 3' from the center line of the posts forming the base. Caps were placed reaching

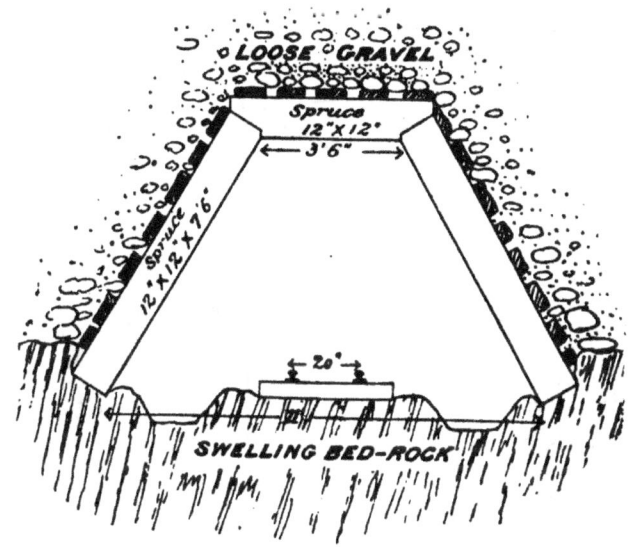

Fig. 22.

from the post at the apex to each of those of the base, and lagging driven in diagonally from the drift. The two sides of the triangular section opposite the base were lagged, a considerable space being left between the lagging to afford an opportunity for the soft swelling ground to force its way through the open spaces, when it is removed. These triangular spaces were continuous; that is, were cut opposite each set of the main drift. The method involved considerable extra expense in mining and timbering, but it was so infinitely superior to any plan previously tried that it was looked upon as a success mechanically and financially. The accompanying sketch (Figs. 23 and 23A) will make plain the details of this peculiar method, which, since its use in the mines on Sugar Loaf Mountain, seems to have been lost sight of.

Drifts sometimes require but a few posts to support particular rock-masses that threaten to fall. Where a post alone affords insufficient support, a heavy piece of plank (plate) is inserted between the top of the post and the roof.

Fig. 24 illustrates a method of timbering in single sets in drift gravel mines. The main gangway is cut into the bedrock somewhat to equalize the inequalities of its surface. The lagging is driven the usual way. The breast is timbered with posts and caps in single (independent) sets, with lagging both close and open, as the nature of the ground may require.

A method of timbering drift mines, using posts and breasting caps, with the employment of a few lagging for good standing ground, is

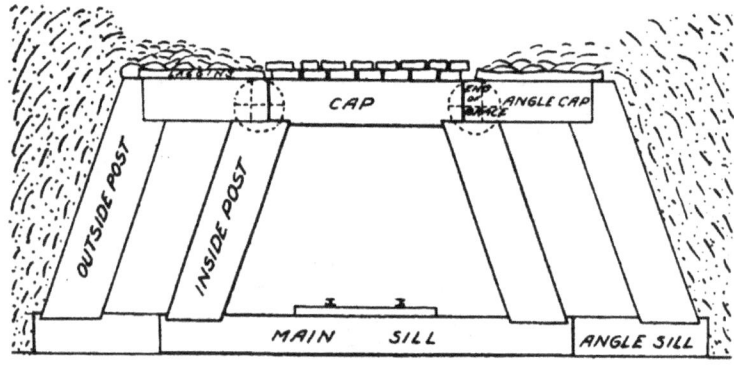

Fig. 23.

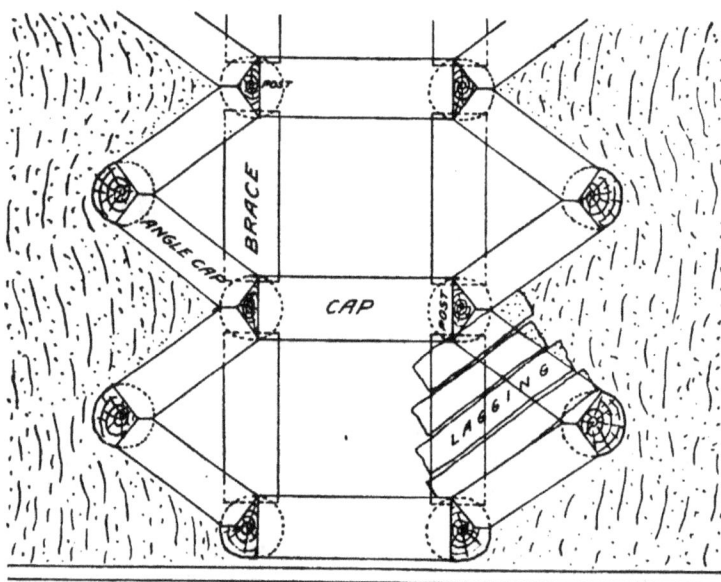

Fig. 23A.

shown in Fig. 25. The gangway is cut into the bedrock, as previously explained. In the breast a platform is arranged at the side of the gangway to facilitate the loading of cars. The manner of setting-up posts and caps, with the distribution of lagging, is clearly shown in the drawing. Should the ground become too loose as work advances, this method must be abandoned, and a system similar to Fig. 24 adopted.

When either of the systems above described are employed, the waste (cobbles, bowlders, etc.) is closely piled in walls behind the workmen, thus supporting the roof better than timbers.

RUNNING GROUND.

Tunnels and shafts must at times pass through soft, running ground. Zones of rock of this description are often found lying between walls of firmer rock. The occurrence is not infrequent on the Mother Lode of California when the fissure is barren of quartz and filled with a mass of soft, crushed, foliated black slate. It is prominent in the Quaker City, Gwin, Hardenburg, Kennedy, and many other mines on the lode. Such ground is nearly always wet, and the process of sinking or drifting in it is attended with expense and danger. In sinking through such ground the miners usually make an effort to push the work and pass through it as quickly as possible. When the ground is very wet and runs easily, it is not always the best plan to "crowd" it. In some cases the difficulties are more easily overcome, the expense reduced, and the completion of the task sooner accomplished by going slowly, allowing the ground to assume a more normal condition by cutting out and removing the material falling into or forced into the excavation. When water is troublesome in ground of this character, the better plan is to permit it to drain off. By doing so the ground sometimes becomes firmer, and, as a consequence, is more easily handled, timbers are more readily placed in position, and the work is carried on more satisfactorily.

Drifting in running ground, through caved workings, or on swelling bedrock is a difficult and often dangerous undertaking, and many methods of timbering such excavations have been employed. Fig. 26 illustrates a method successfully used in Sierra County by Richard Rowlands, M.E., of Placerville. The posts are not set directly upon the rock, but upon a wedge-shaped block, which rests upon two sets of foot-blocks placed at right angles, one above the other. Sills are disadvantageous in swelling ground, but these foot-blocks give a firm support to the post while not offering too great resistance to the swelling ground. The wedge-blocks are intended to prevent the posts from taking a greater spread. Should the bedrock or floor of the drift be firm and hard, and the loose or running material be found only overhead and at the sides, the foot-blocks and the wedges may be dispensed with, and the posts may be given a somewhat lesser spread. It will be noticed that lagging is driven at the sides as well as overhead, and that the bridge and block "A" are in use in each case. A novel feature is found in the framing at the inside corner of post and cap. The former is framed on a bevel, the latter at a right angle. The system contemplates the use of "false sets" while drifting, and the introduction of line (intermediate) sets at some subsequent time, should the nature of the ground require it. Face (breast) boards and head-blocks are also employed. For the former, 2" plank will usually suffice. They may, for convenience, be of various widths from 6" to 12"; and reach, as shown in the figure, entirely across the drift. Head-blocks are made from 4" lumber, and 6" to 12" wide, to correspond to the width of the face-boards, and are sawed off in 6" lengths. They are inserted between the forward end of the lagging and the face-boards. The excavation proceeds by cutting away the earth from beneath and behind the face-board, which is kept advancing by driving up the lagging, working from the bottom upward; or, if permissible, the boards may be removed entirely, one at a time, the earth cut away as far as possible, and the board replaced before the next is removed. In this manner the drift is carried forward, with the help of

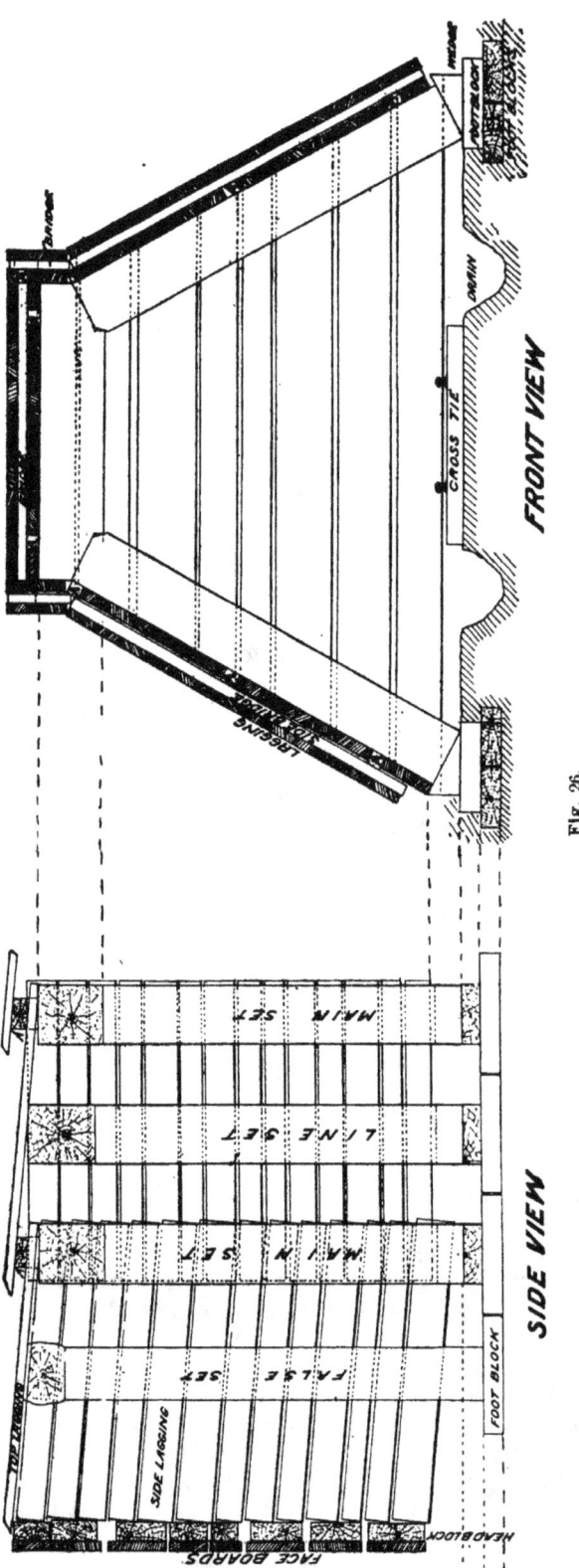
Fig. 26.

the false set, until the full length of the lagging has been reached, when the posts and cap, with the bridges and blocks of the main set, are placed in position. The forward ends of the lagging then rest upon the bridges, and are flush with the farther edge of the set. Excavation is then commenced beneath and back of the lowest face-board, the head-blocks knocked out, the board advanced, and the head-blocks replaced with reference to the new lagging, which is inserted under the bridge and driven forward as before. The second board is advanced in like manner, and so on to the top of the drift.

Another kind of ground difficult to timber is found in some much altered rocks where talc, steatite (soapstone), and serpentine, containing much water, have to be passed through. The ground often breaks well and sometimes stands well for a time, but it is treacherous, and should be promptly and substantially timbered. Rocks of this description are abundant in the great auriferous belt of California, and, as miners there well know, are usually fissured in every direction, and upon exposure to the atmosphere exhibit a tendency to break up (called blocky ground). Great angular blocks and "heads" (round bowlders) drop from the roof and sides without warning. The former are often wedge-shaped, and slip out from the fissured rocks when the ground had appeared firm and solid. The "heads" are usually hard in the center, while the outer portion is quite soft, feeling greasy to the touch. In size these masses range from an inch or two in diameter to those which weigh tons. Timber must be placed in ground like this immediately as the work progresses. The sets should not be more than $2\frac{1}{2}'$ from center to center, instead of the usual distance, 4' to 5', whether it be in shaft or tunnel. Lagging, when used, must also be short.

When passing through slips or fissures, whether single or in zones, in any kind of rock, extraordinary precautions should be taken, as accident is much more likely to occur at such points than in solid, unfissured ground. Rock in the vicinity of veins is nearly always more dangerous than that at a distance from the vein.

SHAFTS.

Working-shafts, as well as tunnels and drifts, should be arranged with a view to securing their permanency. Indeed, in consideration of future possibilities, even greater care should be exercised in the selection of their location and in deciding upon their size, while the manner of timbering is most important. Working-shafts should be so equipped as to remain open and be in use as long as ore remains in the mine which it will pay to extract. As in all other mine work, the amount of timber required depends largely on the size of the shaft and more particularly on the character of the rock through which it passes. Prospecting-shafts are sometimes sunk in good-standing ground to the depth of several hundred feet without other timber than a few stulls, to which ladders are secured. The few timbers thus placed often become insecure through neglect, particularly in regions where there are climatic alternations of wet and dry. When wet, the timbers, and the wedges securing them, swell. With the change to dryness they shrink and are likely to drop out. An additional danger results when the rock walls crumble, and men working below are in constant danger from falling rocks and timbers. The wedges demand frequent attention, for they must be kept driven well in at all times. On the desert and in mines above the timber line, where timber is expensive, miners endeavor to get along with as little as possible, and are not very particular as to the kind and quality of that which is used. It would perhaps be a better plan to dispense with timber altogether than to place too much dependence on sticks that are likely to drop out of position unexpectedly. As a matter of fact, the writer has seen shafts in the Mojave Desert mines more than 200 feet in depth without a single stick of timber. The necessities of the case in sparsely timbered regions and on the desert have obliged the miner to resort to many novel plans to protect himself against danger

Fig. 27.

at the least possible expense. He puts in as substantial a frame of timbers as he can obtain, or as he may think he can afford, using a few frail saplings, thin split lagging, or even brush, to support the sides of his shaft. Fortunately for him, in the desert regions, where scarcity of timber forces upon him this economical "system" of timbering, the rock, being nearly always dry, stands fairly well, as a rule, and expensive timbering is not necessary.

The extremities to which prospectors are often reduced to procure timber in these timberless regions has resulted in the adoption of a method peculiar to such districts. While the result is not particularly pleasing from a workmanlike standpoint, it nevertheless exemplifies most faithfully those principles which are the foundation of the most elaborate system of timbering. In these shafts all timbers are stulls, each one being placed to support some particular rock-mass which seemingly threatens to fall. Each stick is independent of the others; there are no superfluous timbers, and no attempt at system or regularity. As a result, these sticks cross the shaft at many angles. Some are horizontal, but most of them inclined somewhat from that position. It sometimes gives the shaft timbers a spiral appearance as viewed from above. Despite the fact that these timbers are placed so much at variance with recognized methods, if placed at the time of making the excavation, or shortly thereafter, and properly and firmly wedged, they usually render the shaft fairly safe. In that region old redwood railroad ties are very frequently used for mine timbers, no other timber being available.

Shafts having a single compartment, such as are frequently seen in small mines, are timbered in a simple manner. The timbers consist of two wall and two end plates, and four posts to each set. The method shown in Fig. 27 is quite common, and suited to shafts of moderate size (5′ x 7′ in the clear), having a single compartment. The four frame timbers are placed in position and tightly wedged, the posts being driven in at the corners. Care must be taken to keep the sets in line. Sets are ordinarily 5′ apart from center to center. When the ground is

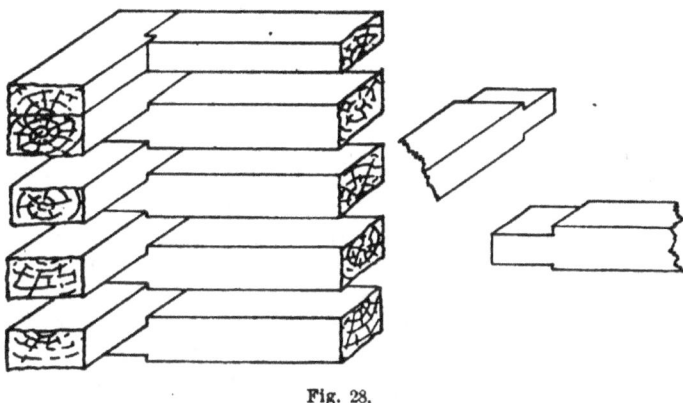

Fig. 28.

heavy, sets may be placed closer than 5′. They are often only half that distance from center to center.

Lagging, either split or sawed (2″ plank), is driven in behind the timbers, the forward end being kept well into the wall by the insertion of a block between the upper end of the lagging and that of the set above. Sometimes a horizontal piece, 4″ in thickness, and reaching from post to post, is used instead of single blocks. This plan keeps the lagging pointed evenly. It is driven one half to three quarters of the way down; then the lower ends of the next set below are inserted between the timber and the wall. Later, the lagging of the set above is driven down to its proper place.

CRIBBED SHAFT.

When the pressure of the ground is very heavy a crib of timbers is built, the timbers being placed one on another, with only a notch at the ends to hold them in place. In soft ground it may be necessary to use lagging, even in a cribbed shaft. All open spaces between the walls and the crib or lagging should be filled with broken rock, to secure firmness, and to counteract any tendency of the timbers to shift. The manner of framing crib timbers is shown in Fig. 28.

METHODS OF MINE-TIMBERING.

REACHERS IN SHAFTS.

When sinking can be carried on somewhat in advance of timbering, it is sometimes the custom, in firm ground, to place long timbers, called "reachers," across the shaft, the ends resting in niches cut into the walls. These having been firmly placed, the four timbers of the set are laid upon them and firmly wedged, and from this foundation the sets are built upward to the next set of reachers above, a distance of 25' to 30'. Where the shaft is sunk in country rock, or in a large pillar of ore (the latter to be avoided when possible), the reachers are placed alternately in sets at right angles.

The manner of framing the ends of shaft timbers where they join at the corners is shown in the cuts of the Requa, Forman, and Alma shafts and the Argonaut incline.

Men placing timbers in shafts usually work suspended on slings secured to timbers above.

IRON DOGS AND BOLTS.

In places where the above described methods of building sets of shaft timbers on reachers is not possible, owing to the soft nature of the rock in which the shaft is sunk, or where it is desirable to run cages to the bottom of the mine, it was formerly the custom to suspend the shaft timbers by ropes in a position as near to that desired as possible, and then to maintain them in that position by driving iron dogs into the timbers, the weight being supported by the set next above, which had previously been secured by wedges. In some cases these dogs were never removed. Many of them may be seen in the older California mines. The dogs were made of round or square iron bars 1"x 1" or 1"x 1½", and having the ends turned at right angles. The points were 3" or 4" in length, and sharp. The length of the dog was determined by the distance from center to center of the sets.

When the rock was sufficiently firm to admit of the timbers being firmly wedged, the dogs were knocked out. Iron dogs or bolts are useless as a means of support when the surrounding rock-masses have once firmly settled on the timbers.

A safer and more convenient device has been introduced in later years, in the form of iron bars having a thread at one end and a ring or hook at the other. These go in pairs, their combined length exceeding by 6" the distance from the top of one set to the bottom of the next set beneath. The manner of using these hangers is as follows: A set having been securely wedged in its proper position, the threaded end of the bolt provided with a hook is passed upward through a hole in the plate (bored in all timbers for the purpose), a washer passed over the end of the bolt, and a threaded nut screwed down onto it. A second "lock-nut" may be used, but is not necessary. A second similar bolt with hook is passed through a hole near the opposite end of the same timber, and also secured with washer and nut. The timber to be placed directly below is suspended by ropes in a position approximating that desired, and a bolt having a ring at one end is passed downward through this timber, immediately below that in the timber above, and a similar bolt is passed through the opposite end. Washers and nuts are placed on the ends of these bolts, and the rings are caught in the hooks of the bolts above. The nuts may now be turned and the timber drawn to exactly the position required, when the ropes may be moved. All four of the timbers of this set having been placed in position, and being suspended on the hangers, the posts are slipped in, the nuts tightened still further, and the whole firmly wedged, when the next set below may be put in in like manner. The hangers are left a few days or weeks, and in some instances permanently; but, as previously stated, they are useless as a means of support when the weight of the surrounding rock has settled on the timbers.

Where, owing to the soft nature of the ground, it is thought desirable to have the hangers remain in place indefinitely, a bar having a thread at one end and simply turned at a right angle at the other, or having a solid hammered head with a washer, may be used, being passed upward through the bottom timber and then through that above, where the adjustment is made by means of a washer and nut; but as these bars are longer than the distance between sets, their removal is impossible, a set below once having been put in place. There is no question as to the superiority of the bolts having rings and hooks for either temporary or permanent use.

RETIMBERING SHAFTS.

It very frequently occurs that shaft timbers have to be removed and new ones inserted, and also often necessary to reinforce timbers already in place. This work frequently necessitates the suspension of hoisting through the shaft. The Superintendent of the Wildman Mine, at Sutter Creek, Amador County, California, having occasion to retimber the shaft between the 500 and 600 levels of the mine, constructed a chute 1' square of 2" plank in one corner of a compartment of the shaft reaching between these levels. It was built in short sections, each the length of a set of timbers. The work of retimbering was commenced above and carried downward, all of the refuse rock, timber, etc., being dumped into the chute. This material was taken up on the 600-foot level, dumped into a skip from time to time, and sent to the surface. By this means the work was quickly and safely carried on, sections of the chute being removed to keep pace with the work; there was very little delay in operating the skips, and hoisting of ore was continued almost without interruption.

SHAFTS HAVING TWO OR MORE COMPARTMENTS.

Large shafts are separated into two or more compartments by placing timbers, called "dividers," at intermediate points between the end plates. They reach from the wall plate on one side to that on the opposite side of the shaft. These dividers are framed with a short, beveled tenon, broader at the top than at the bottom. (See cuts of Forman, Requa, and Alma shafts.) These tenons fit exactly into mortises or notches in the wall plates. At each of the four corners of the shaft and between the wall plates, opposite each divider, is placed a post, which is set in a shallow "dap," or seat cut in the plates. In size the posts may equal that of the plates, or if desired they may be of smaller dimensions. The dividers are made the same depth as the wall plates, but are usually narrower across the upper face.

The dividers separating shafts into two or more compartments may be made most secure by permitting the upper portion of the beveled tenon to extend into the wall plate 2" or 3", the bottom portion being let in but 1". The post setting directly on the divider holds it much more firmly, and the danger of having the dividers knocked out by shots fired directly below is greatly decreased, if not obviated entirely. It not infrequently occurs, when dividers are framed so as to set but 1" into the wall plates, that a heavy blast will tear them out altogether, incurring expense of timber and time, which may be avoided in the manner stated above. The large new shaft of the Calumet and Hecla mines in the Lake Superior copper region is rectangular, and has three double compartments, formed by a dividing plate passing through its center.

Drawings of two Comstock shafts illustrate the manner of framing and placing timbers in them. One of these is called an "L" shaft, and was sunk by the Overman and Caledonia companies jointly. It is known as the Forman shaft. The other is the Requa shaft, sunk by

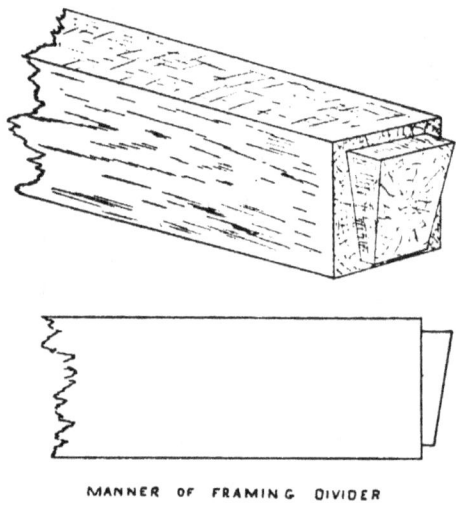

MANNER OF FRAMING DIVIDER

Fig. 29.

the Chollar, Norcross, and Savage companies. It is a large rectangular shaft, having four compartments, and is a splendid example of its kind. An illustration of the new Alma shaft at Jackson, Amador County, California, shows a different style of framing. In fact, the manner of framing and joining the timbers of these three shafts is totally different.

In size, shaft timbers range from 8" x 8" to 20" x 24", and even larger dimensions are sometimes employed in very heavy ground, particularly in inclined shafts. The wall plates are usually broader on one side than on the other, and are placed in position with broad side up. End plates are usually the square of the smaller dimension of the wall plate.

Pumping and manway compartments do not require lining, but hoisting compartments, particularly where cages are run, should be lined throughout, to prevent accident to men who sometimes overcrowd a cage. There appears to be less danger of this where buckets or skips are used. Every shaft in a mine should be provided with ladderways as a means of exit in case of accident to shaft or hoisting machinery.

Sinking large shafts in swelling ground, in loose, watery, running ground, or in quicksand greatly multiplies the difficulties and dangers

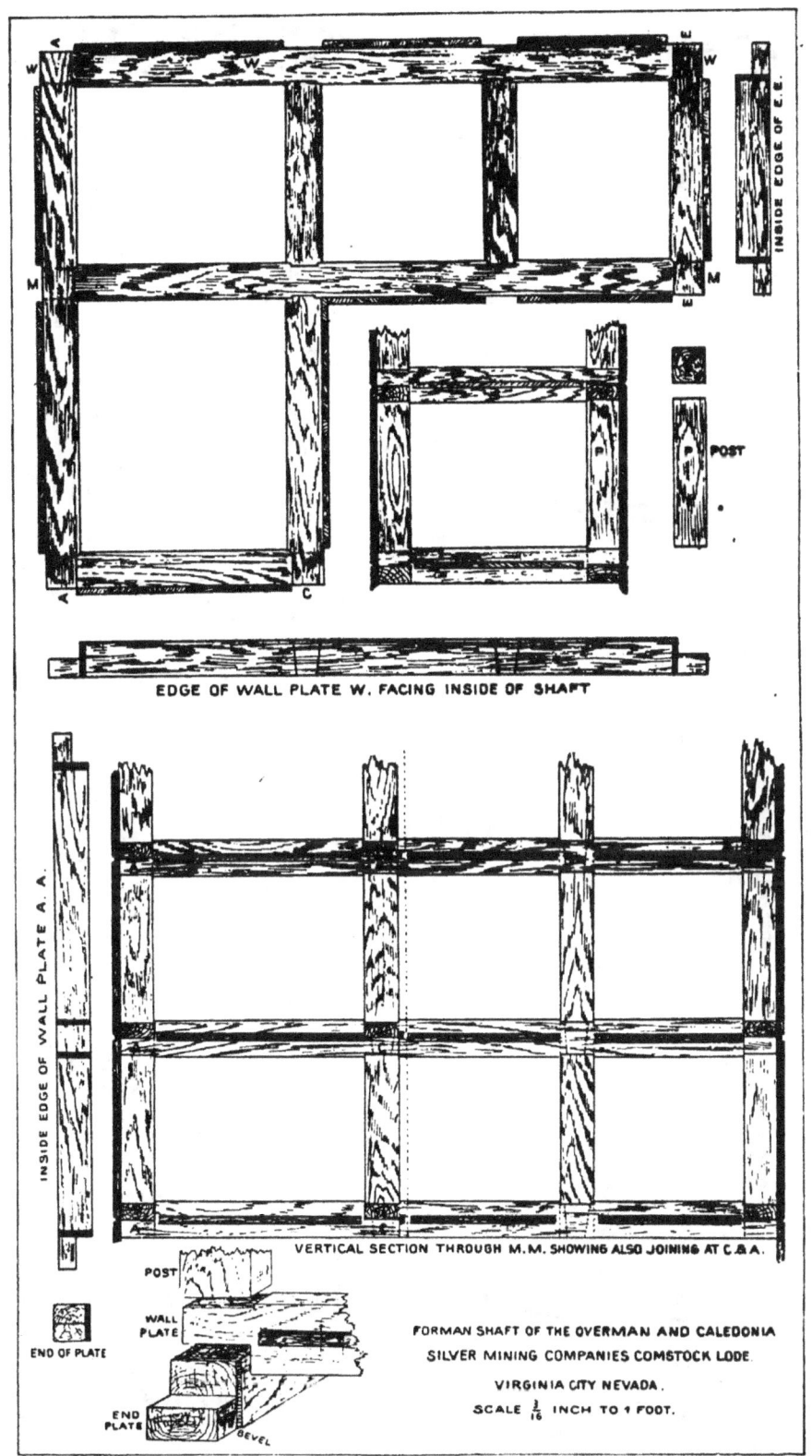

Fig. 30.

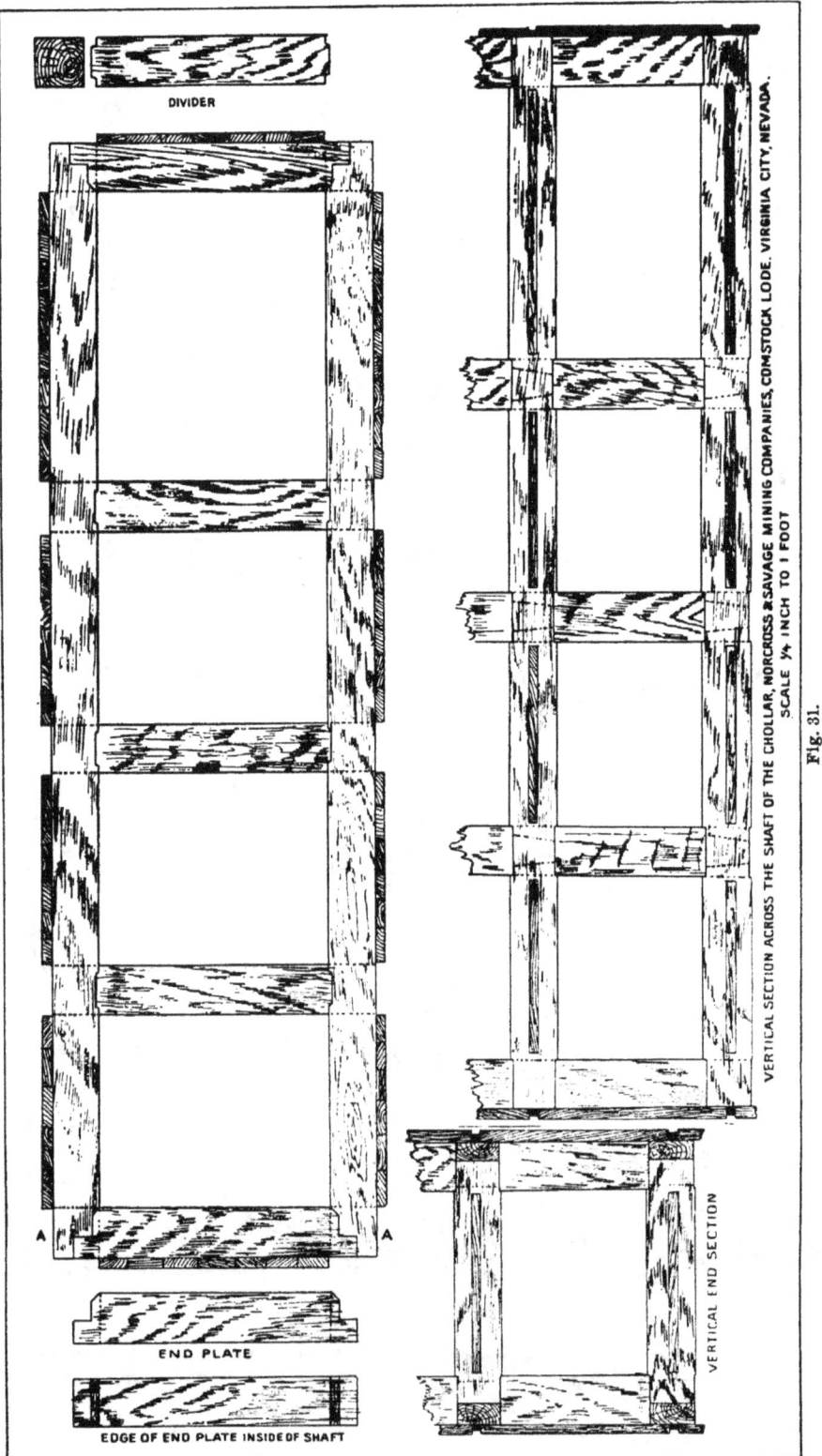

of the miner. There are instances where shafts have been abandoned owing to insurmountable obstacles which a combination of engineering skill, capital, and labor was unable to overcome; but such instances are of rare occurrence. Few shafts present greater difficulties than were

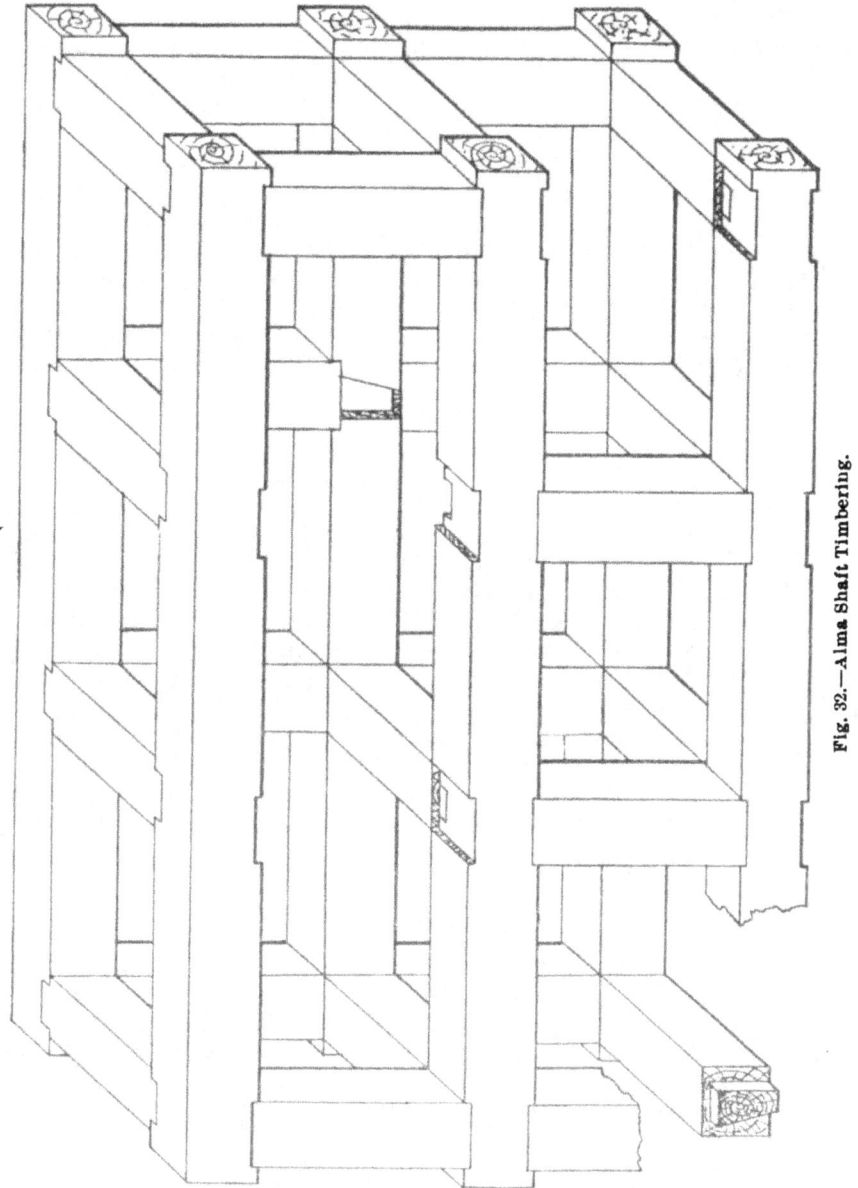

Fig. 32.—Alma Shaft Timbering.

encountered at Leadville, Colorado; in the Lake Superior region; in some coal mines; on the Comstock, and in Calaveras and Amador Counties, California (the latter are mostly inclines). Some of the shafts in the regions mentioned are to-day splendid and substantial monuments to American engineering and to the enterprise of the companies owning them.

The shaft of El Capitan Mine, Nevada County, California, now the property of the Providence Mining Company, was sunk through decomposed slate and altered dyke rock under the most trying and dangerous conditions. After several failures, the shaft was carried down to solid ground under the direction of Francis Burns, who worked slowly and carefully, giving the soft, running ground time to drain. The water being pumped from the sump, after standing for a time, appeared to drain the ground, and it was found that by this means progress was much more rapid than when the work was hurried. "Breast-boards" were carried down until the dangerous places had been safely passed.

It is sometimes necessary in sinking large shafts to carry them down in sections by driving lagging or planks down in advance of the excavation. This is known as "fore-poling." This section is started at one end of the main opening and advanced several feet, the lagging being well braced. Considerable water will drain into this depression, making the removal of the ground from the remainder of the shaft much easier. By carrying one side of the shaft somewhat in advance of the remainder, loose, wet ground can be worked more quickly and at less expense than when it is attempted to carry down the entire shaft at once.

LAGGING IN SHAFTS.

In large shafts sunk in fairly good ground, where it is not necessary to drive the lagging nor to "fore-pole," the lagging is easily placed in position. It is usually made from 2" plank, but in very heavy ground from 3" plank. It is cut 56" long where the sets are 5' from center to center, the usual distance. Before the wall plates are lowered into the mine, wooden strips 2" square are securely nailed to the center of the outside face. When the timbers are in position, the lagging is slipped in diagonally behind the plates and then placed upright, the lower end resting upon the 2" strip secured to the lower plate. One by one the several pieces are put in place and the space behind them filled with blocks of wood, loose rock, etc. The last one or two of the lagging on a side are inserted with the assistance of a hand dog driven into their inner face. When secure, the dog may be withdrawn. (See Comstock shafts, Figs. 30, 31.)

SPLICED WALL PLATES.

In shafts where sinking cannot be carried on much in advance of timbering, and where dividers must be placed in position at once, owing to heavy ground, the wall plates may be spliced. The better plan is to extend the plate from one side entirely across the two hoisting compartments, that portion in the pumping compartment being in a single piece, and joining the longer piece directly opposite the divider next to the pumping compartment. These joining ends of the two sections of wall plate should be framed so as to permit the divider to lap over their entire width, the upper half of the plate sections being removed so as to admit the upper half of the divider, as shown in Fig. 33.

The post standing immediately upon the divider, and covering the entire splice and overlap, prevents the divider from being blown out by heavy shots below. This plan also greatly facilitates the placing of shaft timbers, as there is not that great loss of time in getting the wall

METHODS OF MINE-TIMBERING.

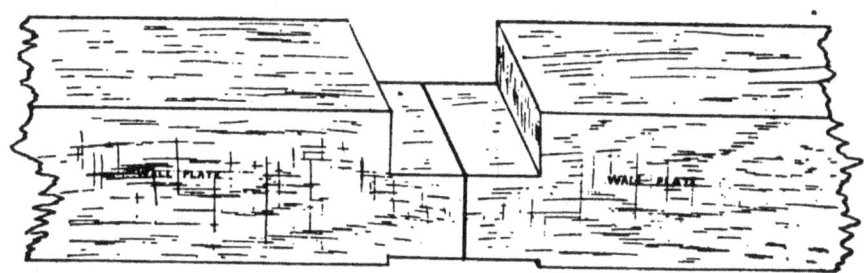

FRAMING FOR SPLICED WALL PLATE AND OVERLAPPING DIVIDER
FOR SHAFTS.
Fig. 33.

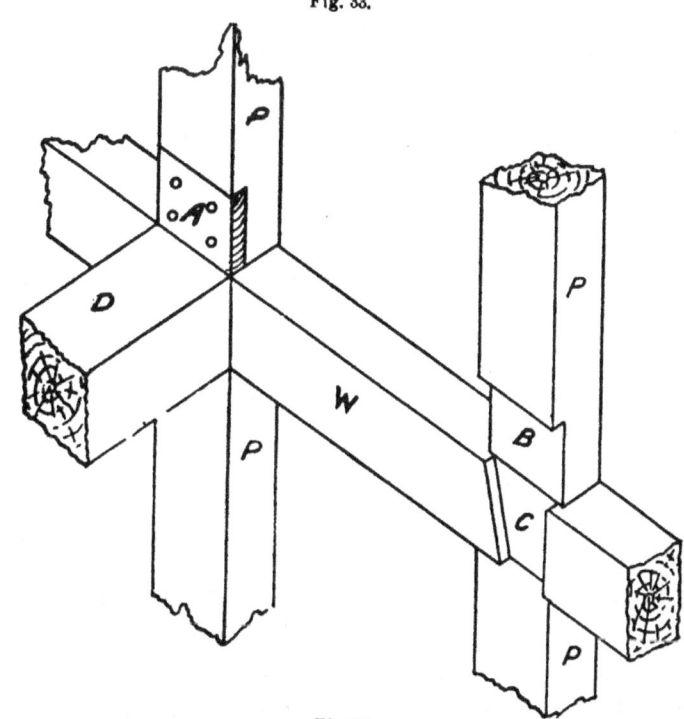

Fig. 34.

plates in position, which results when long timbers are handled. There is always difficulty and loss of time where the shaft cannot be carried down in advance of the timbers far enough to give room to turn the long wall plates. This system also permits of the renewal of wall plates and other shaft timbers by the ease with which they may all be removed.

Fig. 34 shows a Montana method of placing dividers after plates and posts are in position. This method may be conveniently employed where the nature of the ground will admit of leaving the dividers out for three or four sets. This will leave ample room for handling long wall plates without cutting and splicing, as in Fig. 33. A recess, B, is cut on the inner face and lower end of post, P, and temporarily filled with a board, A, which is lightly tacked in place. When ready to place the divider in position, the board, A, is removed and the divider slipped in sideways and allowed to drop into place, the tenon fitting into the beveled mortise, C, of the wall plate, W. The board, A, is then replaced and securely nailed.

BREAST-BOARDS.

When the ground has a tendency to thrust itself up in the bottom of the shaft, the plan of planking and bracing the ground is sometimes resorted to. The same principle is applied in drifting, when the method is referred to as "carrying breast-boards." (See Fig. 26.) In this method lagging is driven down in advance of the sinking, being supported in position above by a frame of timbers, either permanent or temporary. Planks are laid on the bottom of the shaft, being secured by upright posts, which abut against strong stulls or cross-pieces firmly wedged above. When the pressure from below begins to assert itself, a board near the center is removed and the soft, pulpy mass allowed to force itself upward. The material is removed until the ground is eased somewhat, when a second board is removed, the first being replaced somewhat lower than before, and again secured; the material coming up through this new space is removed as in the first instance, and in this manner work proceeds entirely across the excavation. The process is slow and requires considerable timber, but it is a plan which may prove successful when all others fail.

In some instances iron caissons have been sunk outside or inside the timbers to enable miners to pass through very loose, watery ground and quicksand. Where this is not absolutely necessary, wet places have been passed through by making a clay lining between two layers of close planking. It may be from 2" to 12" in thickness, according to the requirements. It has been successfully employed in passing through quicksand and watery, loose ground.

In very bad, wet ground the idea is to form a caisson-like structure with lagging, the interior being sustained, as explained above, by massive timbers, so placed as to resist the pressure of the surrounding rock-mass. The exigencies of each particular case must determine the course to be pursued.

The expedient of freezing soft, wet, running ground, quicksand, etc., is now much resorted to, with great success. It is a patented process, but finds general use where applicable.

Circular shafts are seldom sunk in America now, though very common in the early days in California. There are circular shafts all through the old river-bed region, which pass down through the tufa capping without a single stick of timber from top to bottom. Many of them are over 250' in depth, and, notwithstanding they were made more than forty years ago, are still in good condition and likely to remain so for a century. The rectangular shaft is the most common form in this country. "L" shafts do not find much favor and have nothing to recommend them. The size and number of compartments of a shaft must be determined by the amount of hoisting expected to be done through it.

STATIONS.

Fig. 35 shows the manner of placing timbers at a station in an inclined shaft. The method in perpendicular shafts is essentially the same. The chain blocks are for the purpose of landing timbers sent down in the skip.

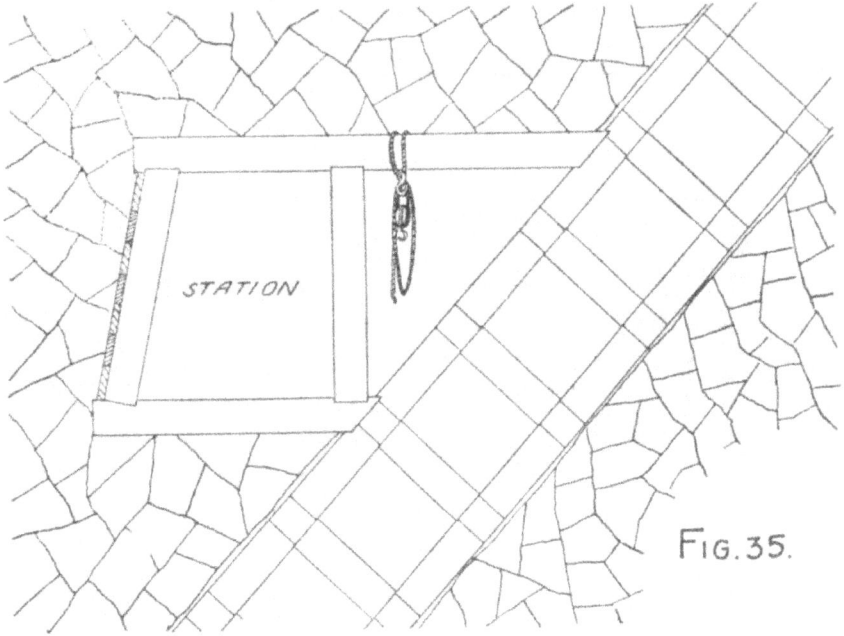

JOINING OF GUIDES IN SHAFTS.

Fig. 36 shows two methods of framing and securing the guides in shafts where cages or skips are used. They should be so joined that there is no likelihood of their warping and projecting beyond their plane, thereby jamming the cage or skip, which is a most dangerous thing and a great strain on the cable. The overlapping of the ends has been for years a common practice; the two ends being secured at the overlap by a single lag bolt. An improvement has been made on this plan, with a view to greater safety, by joining the adjacent ends of sections with a simple tongue and groove, like ordinary flooring or wainscoting. The ends each have a lag bolt, which makes them doubly secure. Spikes or nails should never be used, as it requires much more time to remove a section when repairs are necessary. Guides are most easily repaired by laying a secure platform in the compartment at the point requiring repairs, the men working upward from set to set. Should the whole shaft, or any large portion of it, require new guides at any time, the platform should be put in as described. The hood is removed from the cage and a bottom of loose boards laid on the floor. The workmen at the platform unscrew the lag bolts and take out the old, worn guides. The cage being lowered to the proper point, the old guides are passed up through a hole made in the bottom of the cage, and the new guides passed down in the same manner and put in place. In this way the guides may be replaced for a thousand feet in a single day by a gang of good workmen. Other repairs in the shaft may be made in a somewhat similar manner.

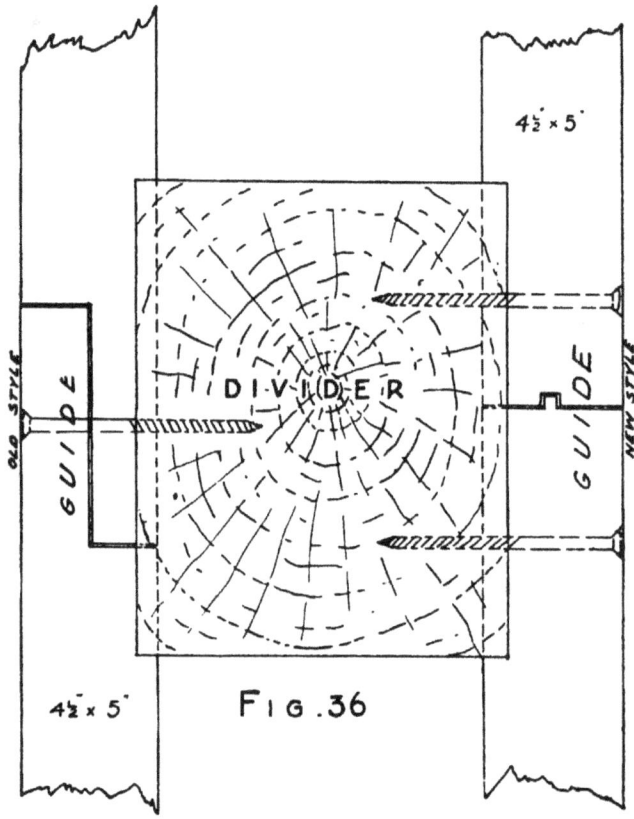

FIG. 36

INCLINES.

Inclined shafts are somewhat different from those that are vertical, and are probably quite as numerous. In California the number of inclined shafts far exceeds those that are vertical, being, as a rule, sunk on the vein, which, in a majority of cases, dips at some angle from the horizontal. In a general way what has been said of vertical shafts applies to those that are inclined. There is considerable difference, however, in the manner of framing timbers for an incline, and while timbers framed for an incline will do, perhaps, equally well in a vertical shaft, those framed, as already shown in the illustrations, are not preferred in an incline. An illustration of the method of framing timbers

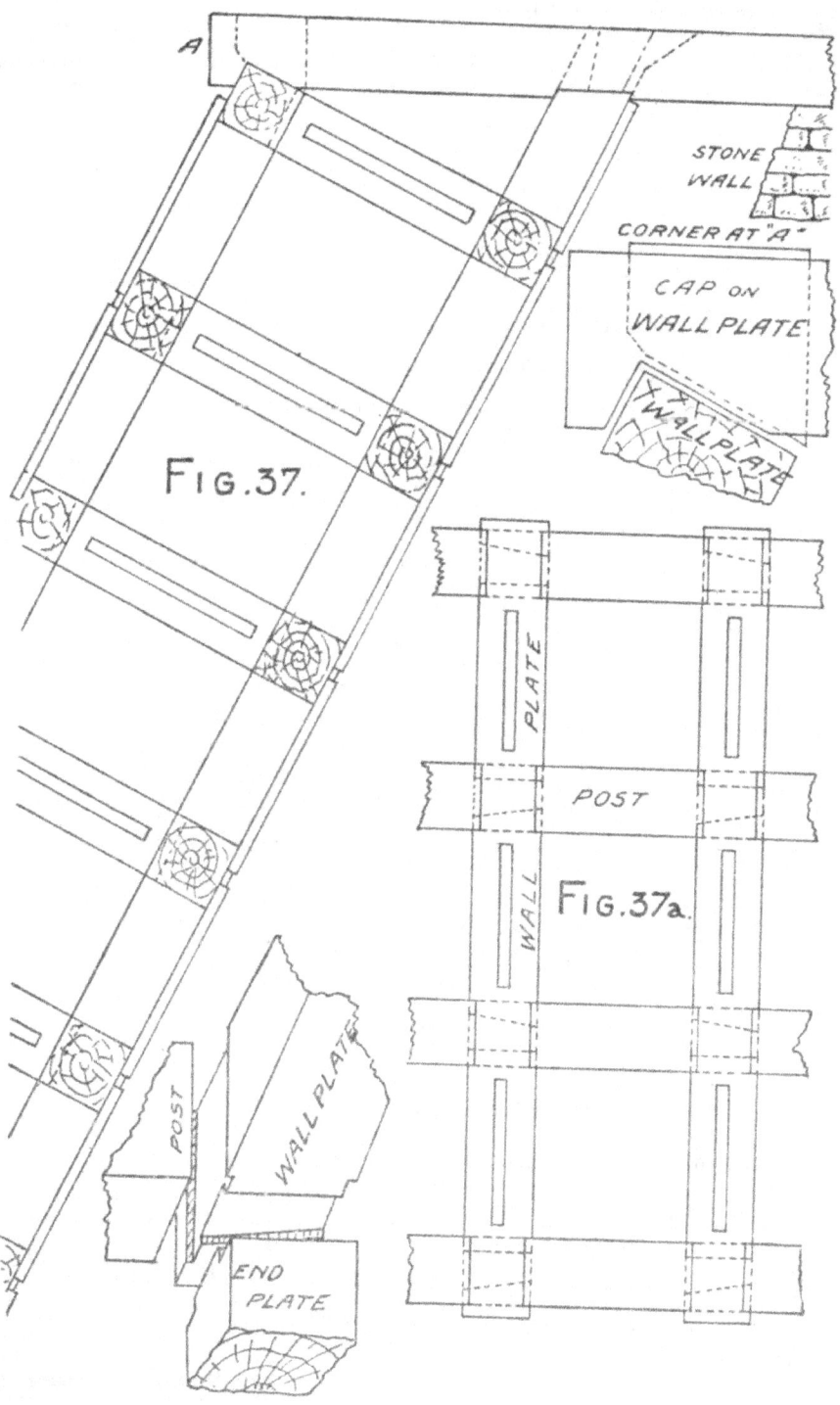

Fig. 37.

Fig. 37a.

for the new Argonaut shaft at Jackson, Amador County, California, which is expected to reach a depth of 2,000' on the incline (63°), is given. It resembles somewhat the new Alma shaft, also at Jackson, though the first 350' of the Alma is vertical.

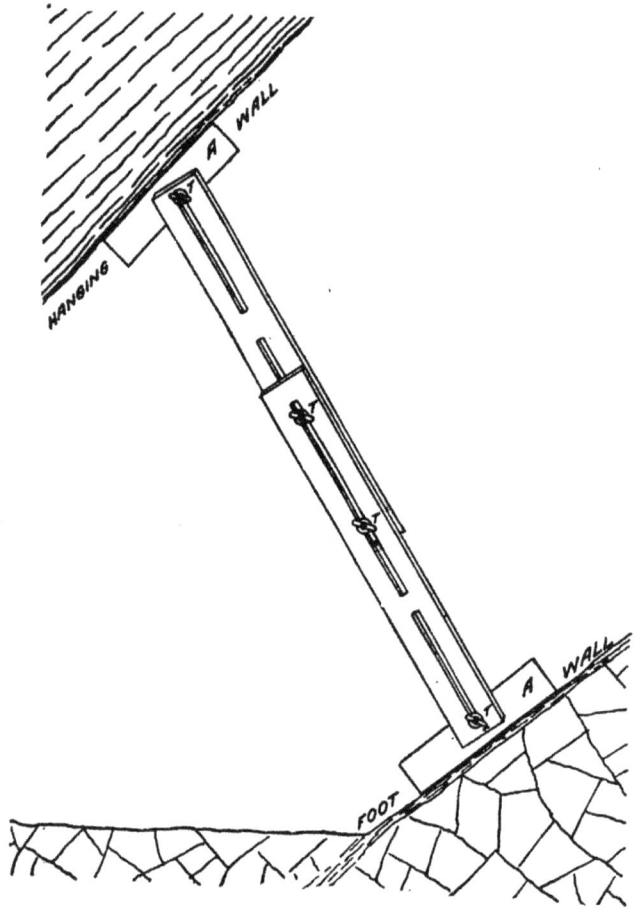

Fig. 38.

STOPES AND CHAMBERS.

The cuttings in mines, which require the most care and greatest skill in placing timbers to support overhanging ground, are "stopes" and large "chambers." The method adopted must always depend upon the size of the excavation, character of the ore and of the walls, the pitch of the vein or ore shoot, and also on the expense of the timber.

TAKING ANGLES AND DISTANCE BETWEEN WALLS.

A very convenient instrument for measuring the distance between the hanging and foot walls of the vein, and for determining their respective angles of inclination, for the purpose of cutting stulls with proper length and bevels, is shown in Fig. 38. It consists of two flat, planed boards, having slots as shown in the drawing, and fitted with thumb-screws, T T T. At either end is a movable arm, A A. These are placed against the walls and the screws tightened, the distance between walls being ascertained by lengthening or shortening the instrument by means of the sliding arrangement. Having carefully taken the required measurements, there can be no mistake as to the kind of stull required.

TIMBERING A SOFT HANGING WALL.

In some portions of the New Almaden Quicksilver Mine, where the width of ore stoped is about 10', an unusual method of timbering has been adopted. (See Fig. 39.) The foot wall is usually hard and solid,

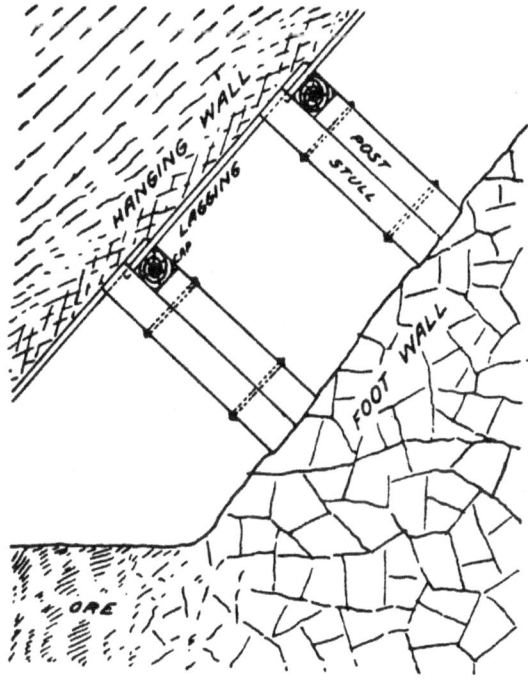

Fig. 39.

while the hanging wall is almost always soft and shelly, and is considered dangerous ground. Heavy stulls are placed at regular distances (about 8'), and are set in line one above another. Immediately above the stull a second shorter timber is laid, which rests upon the foot wall and reaches within about a foot of the hanging wall. A plate or cap is inserted over the upper end of this auxiliary stull, which reaches horizontally across to the next stulls, which are similarly placed. Heavy lagging is driven upward along the hanging wall, the lower ends of which rest upon the cap and the upper ends upon the lower ends of the lagging of the next set above. The two stulls are bolted together near the top and bottom, for additional strength and security. Redwood has been used quite extensively in timbering this mine, and it is claimed to have given satisfaction. The large stopes of the New Almaden are timbered in square sets.

FLAT OR BLANKET VEINS.

In veins which lie quite flat, the thickness of the vein and the system of mining must determine the character of timbering to a great extent, as well as the character of the walls. Where the vein is thin (a foot or two) very little timber is required, the waste rock filling the entire space behind the miner. Where the mineral deposit is thicker and timber is necessary, various methods are pursued. Some ground stands well by simply leaving pillars of mineral. In other cases a series of upright posts and breasting caps will sustain the roof, the posts being placed in rows directly back of the workmen and as close to the face as necessity demands. The foot of the post rests either directly on the rock floor or upon a block of wood or piece of heavy plank. The posts are forced into position by driving them up with heavy hammers. Care must be taken that these posts are so placed as to receive the weight of the roof directly and not at an angle. These timbers are set in lines standing in two, three, or four rows back from the face, the waste being piled behind as the work advances. In this manner, by exercising care, many sticks can be recovered before the weight settles so heavily on the refuse rock as to render it impossible to remove the timbers. Some flat veins make little or no waste. It is then necessary to follow the "pillar and stall" system of extraction, considerable blocks being left to sustain the roof. Posts and caps are used in this system also. Frequently the caps reach in a continuous line from post to post, joining the next set, the ends of two caps resting on a single post; the combined sets being a hundred feet or more in width. Large timbers thus placed will support great weight, but if small rocks fall from the roof lagging must also be employed. This system is much in use in California drift gravel mines. (See also Figs. 24 and 25.)

When a vein lying nearly or quite horizontal, and making no waste, is to be mined, a drift should be run along the lowest portion of the deposit, this point having been reached by incline or shaft. The work advances towards the surface, good sized pillars being left to sustain the roof. If timber be necessary, it is put in place in the manner required. The work having advanced sufficiently far towards the surface, the pillars may now be cut out at the back end, while the work progresses as before. As the pillars are removed more timber must be put in, or waste from the surface must be piled in cribs of timber built in the workings already made. Usually some timber can be recovered in this way, and the caving of the roof after the complete removal of the ore, or mineral, does no harm. The main gangways should be substantially timbered, if necessary, as it is desirable to keep it open to the lowest working level at all times.

The "long wall" system of extracting ore is usually carried from the surface inward, a main gangway having been first driven ahead to a connection with a ventilating shaft, when possible. All the ore is removed at once, the waste being thrown back of the miners, who carry the breast forward with the center considerably in advance of the sides, the excavation being in form somewhat like the letter "A," with the apex forward. The waste is thrown into the center to support the roof, while the side passages permit of a free circulation of air all along the face.

STEEPLY INCLINED VEINS.

In vertical or steeply inclined veins, the principles governing the methods of timbering are essentially the same as those above explained, though the application is different. In such workings the post of the flat vein becomes a "stull."

OVERHAND STOPING.

When a working drift is driven along a vein in ore and it is the intention to "stope" out the ore, the character of the walls and the width of the vein or deposit must determine the method of timbering. If the walls be hard and firm, and the vein not more than 10′ or 12′

Fig. 40.

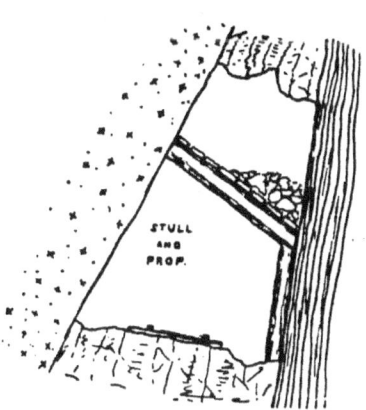
Fig. 41.

wide, posts are not absolutely necessary. Stulls of the proper length are placed with the lower end resting on the foot wall, or in a niche cut for the purpose, the upper end being placed against the hanging wall and driven downward until it stands at a few degrees above a right angle with that wall. (See Figs. 42 and 44.) When the foot wall has a greater inclination than the hanging wall, a support resting on the floor of the drift must be placed under the lower end of the stull. Should the foot wall have a less inclination than the hanging wall, the foot of the stull may be secured by driving in stout wedges from the upper side. (See Fig. 40.) Where the foot wall is sufficiently hard and firm, a niche in either case will answer as a secure rest for the lower end of the stull. Where the vein is more than 10′ in width and additional support appears necessary, resort may be had to the plans shown in Figs. 45 to 50 and 55 to 57.

In case either wall, or both of them, is too soft to safely sustain stulls under the pressure of ore or waste to be piled upon them, a plate (usually a 2″ or 3″ plank) is inserted behind the stull on either wall or both, as may be necessary. (See Figs. 43 and 55.)

When the walls are very soft, and, even with the use of plates, render timbering insecure, the stulls may be placed at right angles to the walls, the ends resting upon posts. Mining upward, the posts of each floor rest upon the stulls under foot and immediately above the posts on the floor below. In this manner a stope in soft walls may be carried to any

38 METHODS OF MINE-TIMBERING.

desired height. (See Figs. 51 and 52.) In such instances it is necessary to place longitudinal braces (ties or sprags), reaching from cap to cap, at right angles to the stull. These ties may be of smaller dimensions than the stull and post timbers.

The method shown in Fig. 45 may be employed where the vein exceeds 8' or 10' and when sticks sufficiently large to support the weight of ore or waste cannot be obtained. Fig. 46 may be employed in the same manner with a greater width of vein. Both of these instances presume considerable waste to result from mining, which will, to a great extent, fill the excavation. Fig. 49 affords a firm support to a stull in a wide place in the vein, and Fig. 48 the same in a still broader vertical

Fig. 51.

Fig. 52.

vein. Fig. 55 shows how, in a vein of low dip and considerable width, a stull may be firmly supported to retain waste or ore coming down from the stope above. In soft vein and wall rock a substantial method of placing sill, stull, and post is shown in Fig. 56.

Stulls are placed at distances ranging from 2½' to 6', as may be required. On them are laid split lagging, forming a floor. As the miner stands on the floor thus improvised, he breaks down the ore, separating it from the waste, if there be any, and sends the ore into the "level" or main passageway below, allowing the waste to accumulate on the floor. The waste often occurs to such an extent that a portion of it also has to be sent to the surface. When this is the case, very little timber is required in the stopes. When, however, the quantity of waste is small, it is often necessary to build a temporary staging upon which to stand until a sufficient space is cut out above to admit of laying a second tier of stulls to sustain the floor. Floors are 6' to 8' apart, and sometimes even more.

A passageway, called a "mill hole," "chute," or "slide," is cut every 30′ to 50′, for the purpose of sending ore to the level below, as the work progresses upward. The distance between these slides is determined by the dip of the vein. They may be a greater distance apart in a steeply inclined vein than in one having a low angle of inclination. Rock will not run freely on a slide having a slope of less than 40°, and more is preferable. In cases where the slope angle is low, it is a good idea to line the slide with plank to facilitate the delivery of the ore to the level beneath. In opening a stope between levels, the best method is to make a raise from level to level, building a loading chute at the bottom. Then at a short distance to one side (15′ or 20′) a second raise is carried up about 15′ and connected with the winze. From this point the stope is opened, the excavation being carried upward and the ore being passed down through the winze to the loading chute. Instead of having two raises as above described, a single raise is often divided into two compartments, one of which is used as a manway, being provided with ladders. As the floors are carried up, a crib is built around the winze, keeping it constantly open. This plan secures economy of labor, and affords the best obtainable ventilation. The stoping of ore continues to within a few feet of the level above, and is then discontinued for the time being, this mass of ore and that remaining within a few feet of the floor beneath being left to be taken out the last thing before abandoning this portion of the mine.

When ore is very rich, it is the custom to blast or pick it down upon canvas or boards, keeping it separate, as far as possible, from the waste, the ore being sacked in the mine. The methods above described apply to overhand stoping; that is, excavating from a level upward.

UNDERHAND STOPING

Is the term used to indicate an operation the reverse of that just described, being the method by which the miner takes out the floor of his level and continues the excavation downward in a series of steps, 7′ or 8′ in height. In this method it is best, for the economical handling of the ore, for ventilation, and for drainage, to have established a connection by winze with the level next below, or it will be necessary to hoist all the ore and water from the stope to the level above. The waste, if there be any, is thrown on platforms or floors between the face of each floor and the winze, slide,* or chute. This method is not advisable except in narrow, rich veins.

Underhand stoping is not commonly followed, but may be recommended in working small veins of very rich ore. It requires usually more timber than overhand stoping, and the timber cannot be recovered as in the latter method. When underhand stoping is in progress with no mill hole to a lower level, or when winzes are to be sunk, much time may be saved in handling ore and waste on the level by arranging some device by means of which the skip or bucket may be raised high enough to dump into a car or into an ore-bin from which a car may subsequently be loaded. This may be accomplished by continuing the skids or trackway above the floor of the level to the necessary height, the windlass or hoist being placed on the farther side of the track or back

---

*The term "ore slide" appears preferable to the word "chute," as conflict is thereby avoided with the word "ore shoot," as applied to a body of ore.

of the winze in a recess cut out for it, and the rope passed over a sheave secured at the proper point overhead. By this means the skip or bucket may be raised higher than the top of the car and dumped automatically or by hand. If the winze be vertical, the same principle may be applied in a number of ways. (See Fig. 53.)

CONNECTING LEVELS.

When stoping by the overhand system, on approaching the floor of the level above it is necessary, where posts and caps have been used,

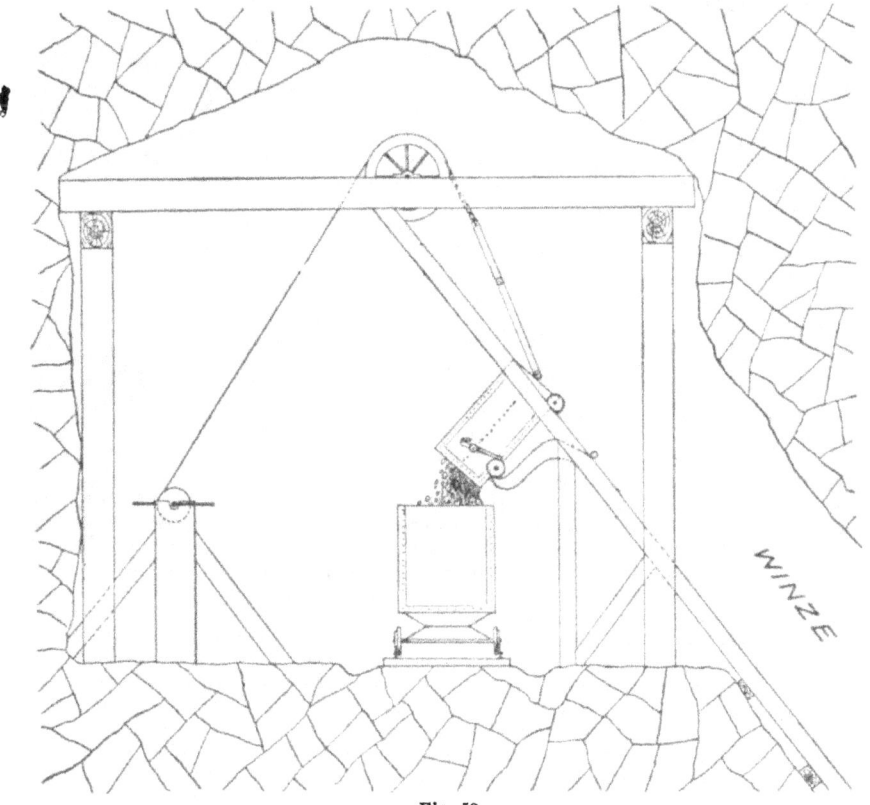

Fig. 53.

whether sills were used or not, to take some precaution to prevent the falling out of these timbers and the consequent caving of the filled stope above, should there be any. The plan adopted in the Bimetallic Mine, near Phillipsburg, Montana, is the most expeditious and the safest. (See Fig. 54.) When ready to break through the floor under any particular sill or set of timbers, a stout piece of timber (sprag) is placed between the posts a few inches above the floor, at A, and tightly wedged with shingles. A heavy stick of timber, B B, long enough to reach across three sets (kept on each level of the mine for this purpose), is lifted to the roof of the gangway, midway between the posts, one end being under the cap which forms a portion of the set of which the sill to be removed is also a part. This timber acts like a lever, having a fulcrum at C, in the form of a post which supports it, the foot resting in the center of the drift on the sill D. Wedges are driven in at E, making it firm and rigid. Wedges are then driven in at F and at C, and the rock beneath may then safely be extracted, the timber-set, G G, being

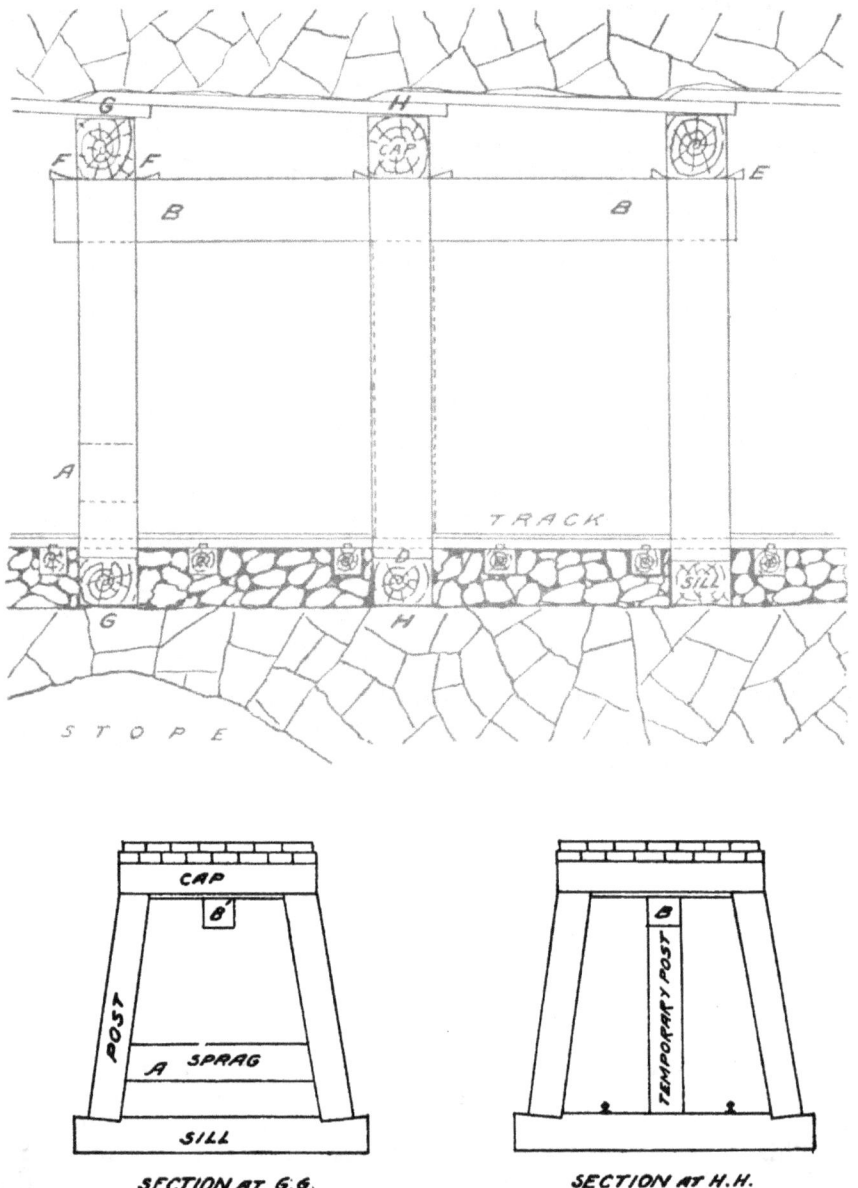

Fig. 54.

held in position, the superincumbent weight being transferred to the points D and E. As a matter of course, the sill of the set G G will drop out when the rock upon which it rested has been removed. The remainder of the set may then be connected with the timbers of the sets beneath in any manner the case requires.

METHODS OF MINE-TIMBERING. 43

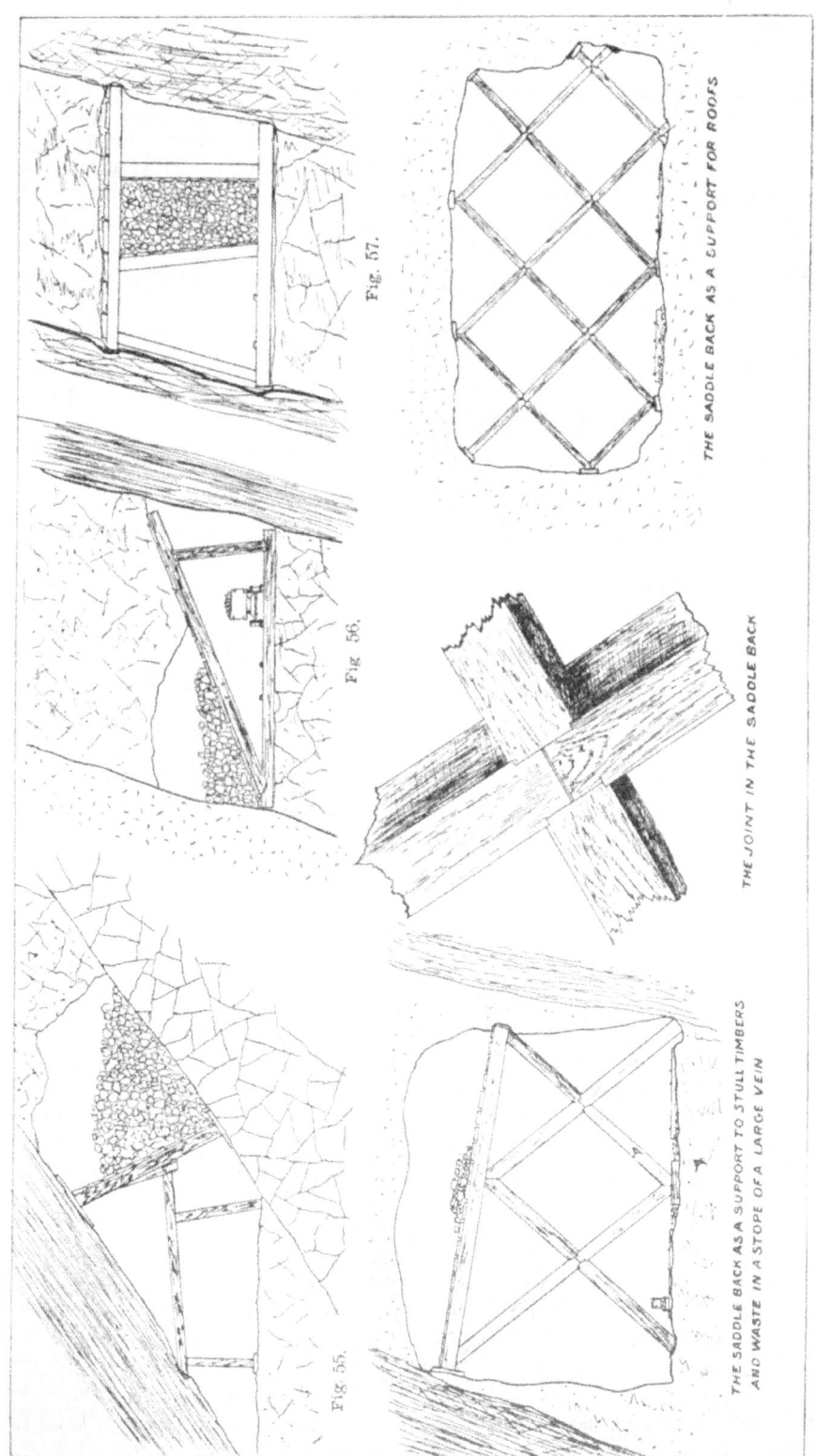

Fig. 55. THE SADDLE BACK AS A SUPPORT TO STULL TIMBERS AND WASTE IN A STOPE OF A LARGE VEIN

Fig. 56. THE JOINT IN THE SADDLE BACK

Fig. 57. THE SADDLE BACK AS A SUPPORT FOR ROOFS

STRENGTH OF TIMBERS.

In placing stulls to sustain the weight of waste, or to store ore already broken, the size and number of timbers to be employed must be determined by the width of the vein and the height to which the waste or ore is likely to accumulate. It is considered better to increase the number rather than the size of stulls. With good walls, stulls 7' in length, having a thickness of 12", placed 30" apart, are calculated to sustain 60' of waste or broken ore in a vein standing at a high angle. After waste has remained in place for some time it settles, and, in some instances, becomes so firm as to retain its solidity after the stulls have fallen out or have been removed. This need not be expected in wide veins.

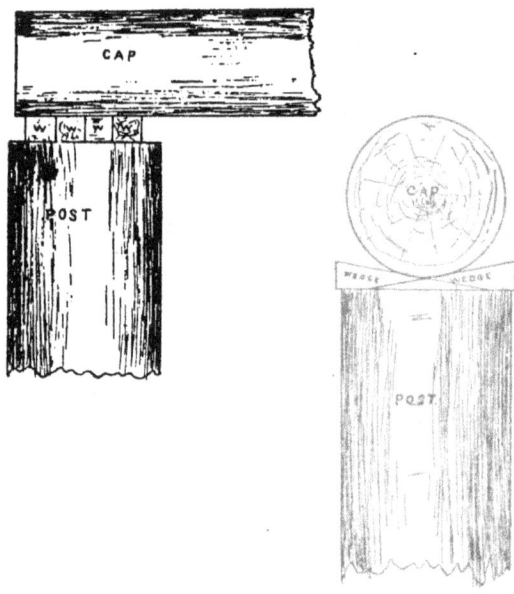

Fig. 58.

When it becomes necessary to quickly catch up settling stull-timbers or caps, a stout post, which will just slip under the sinking timber, is placed in position, and four stout wooden wedges, having the inclined planes sloping alternately in different directions, are driven in between the post and stull. This plan will recover the subsidence, if not too great, and permit of reinforcing the timbers without losing ground. (See Fig. 58.)

SADDLE-BACKS.

A peculiar method of timbering, known as the "saddle-back," is in vogue in some portions of Colorado. It appears to do well in the lead-silver mines having good limestone roofs and walls. It may be considered a modification of the "square set." It requires less timber and is far less substantial. It will not do in heavy ground. The details differ with the various conditions encountered, but the principles are those obtaining in all other timber systems—the application of resistance to pressure. The drawings illustrate one method of setting-up the saddle-back. (See illustrations on page 43.)

Fig. 59.

LOADING CHUTES.

Fig. 59 shows a design for loading chute or slide, though it is not always made as elaborate as represented in the drawing. The inclination of the bottom should exceed 30°, as ores that are soft and wet require the slide to have a greater pitch than those that are dry. Between the two upright posts which reach from the floor of the drift to the roof or to stout stulls overhead, two short uprights are placed, and on top of them, reaching from one to the other, is laid a 4"x 6" scantling. The gate, furnished with ratchet and wheel and pinion, may be dispensed with, loose boards being substituted, which may be pried up, when desired, with a bar. It is well always to build loading chutes in a substantial manner, so that rebuilding may not be necessary. It is a wise plan to provide false sides and bottom, which may be quickly renewed.

The posts supporting loading chutes are not always placed perpendicular. In steeply inclined veins it is sometimes desirable to set them inclining forward. In vertical veins they are built across the vein.

In square sets, ore-bins are constructed within the sets and the chutes are attached to the caps or ties, being arranged at convenient distances from each other.

In some mines, as in the Golden Gate, at Sonora, in Tuolumne County, California, large storage bins are constructed below the floors of the levels at the hoisting shaft provided with a loading chute. The skip is stopped at the proper point and loaded. In this manner the necessity of waiting for cars, or loaded cars and men waiting for the skip, is obviated. In this mine loading chutes are provided on the several levels for the stopes overhead, as a matter of course being constructed in the ordinary manner.

TANKS.

The construction of tanks of large size, as well as ore-bins, at stations under the floor of levels, is recommended in wet mines, as large quantities of water may be prevented from going to greater depths, thus saving expense in pumping. In some California mines, tanks located on the upper levels catch the surface water, and the bottom of the mine, 1,000' or more below, is comparatively dry.

GREAT CHAMBERS AND SQUARE SETS.

The systems of timbering hereinbefore described refer particularly to veins having a width not exceeding 12', though mines have been worked under great difficulties, the operations being attended with extreme danger, where the distance between walls was 20' and even 25'. An instance may be mentioned in California in the Mount Jefferson Mine, at Groveland, in Tuolumne County, where the vein was 25' from wall to wall. A very ingenious system of timbering was introduced in 1854, or thereabouts, consisting of long stulls supported by wall and inside props 7' apart. Longitudinal braces, or ties, were also introduced to support the timbers longitudinally, but the support was insufficient, and a most disastrous cave resulted. It is a matter of absolute impossibility, however, to recover, by the methods thus far given, all the ore from such great masses of mineral as were found in the Comstock Lode, where one ore-body, the Great Bonanza, measured 340' in width at one point, 600' feet in height, and 1,250' in length. Stopes in the various mines of the Homestake group, in the Black Hills, South Dakota, range from 40' to 150' in width and several hundred feet in length and height. The Caledonia Mine, of this group, measured, on the 300-foot level, 195' horizontally. The Homestake vein at the surface, in the open cut, is 360' wide, by actual measurement.

In California there are many mines of great value that cannot be worked by any system of stulls. The Stonewall Mine, in San Diego County, has 20' or more of vein in places. The Odessa, Occidental, Oriental, Silver Monument, and Waterloo Mines, of the Calico District, measure 30' to over 100' in width. Some of the mines of Bodie have very broad veins. The Josephine Mine, in Mariposa County, has an immense ore shoot 50' wide. The Utica-Stickle Mine, at Angels, Calaveras County, is 40' to over 100' wide, and the Gover Mine, of Amador, has an ore-body 30' to 50' wide. The Zeila, at Jackson, is working an immense mass of ore. The Boston Mine, near Mokelumne Hill, is 40' to 60' wide, and some of the ore-bodies in the larger quicksilver mines are of prodigious size. In addition to these there are many other mines in the gold belt of California where the great width of vein precludes the extraction of the ore by the use of any system of stulls or simple posts and caps.

Veins and ore-bodies of large size can be safely, completely, and cheaply mined by using what is known as the "square-set" system of timbering, introduced in 1860, by Philip Deidesheimer, while Superintendent of the Ophir Mine, on the Comstock Lode. It is imperative, however, that stopes be filled as the work progresses, to insure safety.

INVENTION OF SQUARE SETS.

The following interesting reference to early mining on the Comstock is from Monograph IV, of the United States Geological Survey, "Comstock Mining and Miners":

"At the 50-foot level (of the Ophir Mine) the vein of black sulphurets was only 3' or 4' thick, and could readily be extracted through a drift along its line, propping up the walls and roof, when necessary, by simple uprights and caps. As the ledge descended the sulphuret vein grew broader, until at a depth of 175' it was 65' in width, and the miners were at a loss how to proceed, for the ore was so soft and crumbling that pillars could not be left to support the roof. They spliced timber together to hold up the caving ground, but these jointed props were too weak and illy supported to withstand the pressure upon them, and were constantly broken and thrown out of place. The dilemma was a curious one. Surrounded by riches they were unable to carry them off.

"The company was at a loss what to do, but finally secured the services of Philip Deidesheimer, of Georgetown, California, who visited and inspected the treasure-lined stopes of the Ophir."

That the ore-body could not be extracted in the usual manner was at once apparent, and Mr. Deidesheimer says he set about his task with some misgivings. He did not at one stride grasp the idea of the square set, but the system which now bears his name was the outgrowth of circumstances and the very necessities of the case. He instituted a policy, however, the wisdom of which soon became apparent.

The first step was to cross-cut the vein from wall to wall, starting from a drift on the hanging wall side of the vein. As the work advanced he set up posts and placed caps above them, not across the course of the drift, as is usually done, but along the sides, the idea being to form, when completed, a line of caps which would reach continuously from wall to wall. To accomplish this the ends of two caps were placed upon each post, except at the ends. These novel sets were held in place by pieces of 2"x 4" scantling 4½' in length and reaching across the drift from near the top of a post to that opposite.

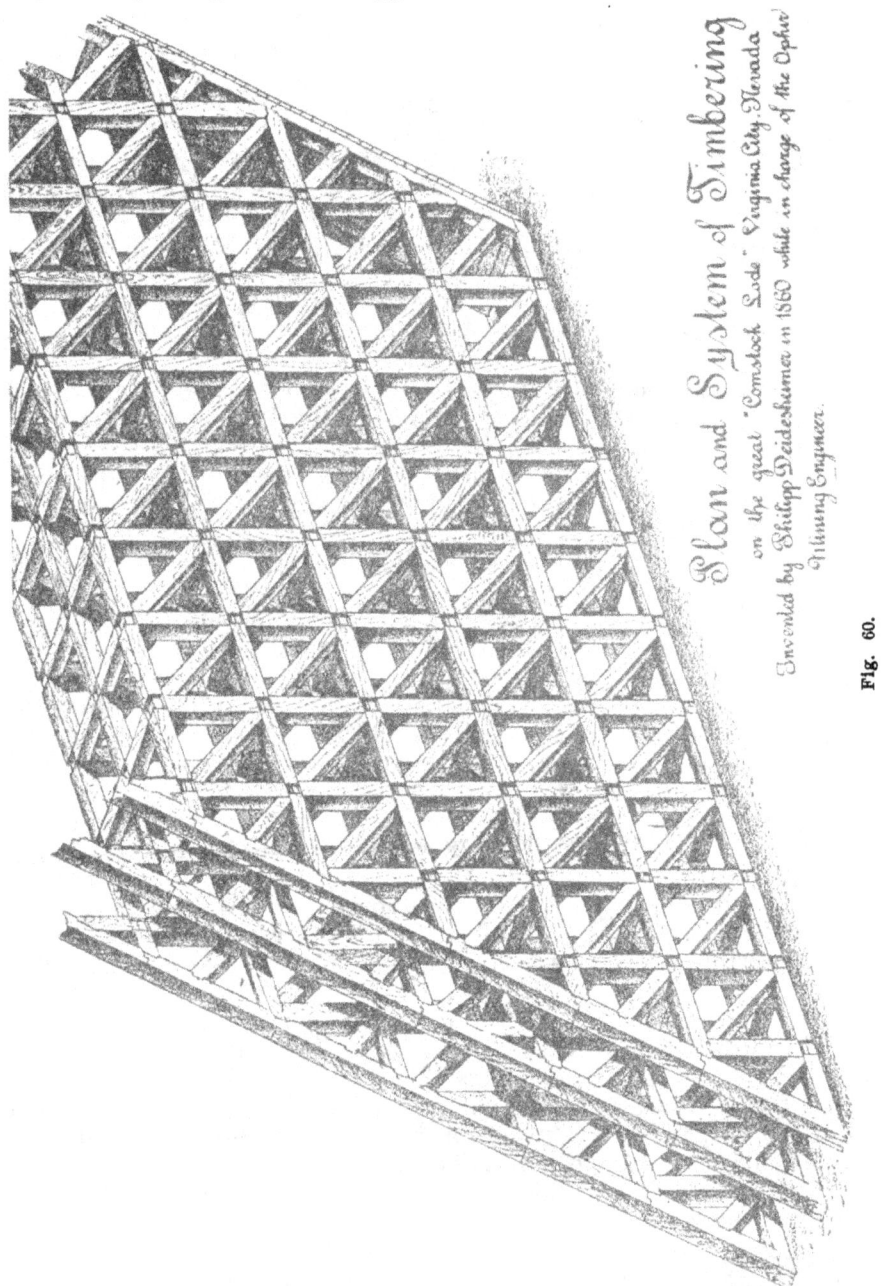

Fig. 60. Plan and System of Timbering on the great "Comstock Lode" Virginia City, Nevada. Invented by Philipp Deidesheimer in 1860 while in charge of the Ophir Mining Company.

Having successfully driven the cross-cut, Mr. Deidesheimer now ran a drift some distance along the foot wall, timbering with posts and caps in the ordinary manner; that is, the caps were placed upon the posts at right angles to the drift and parallel with those in the cross-cut. The

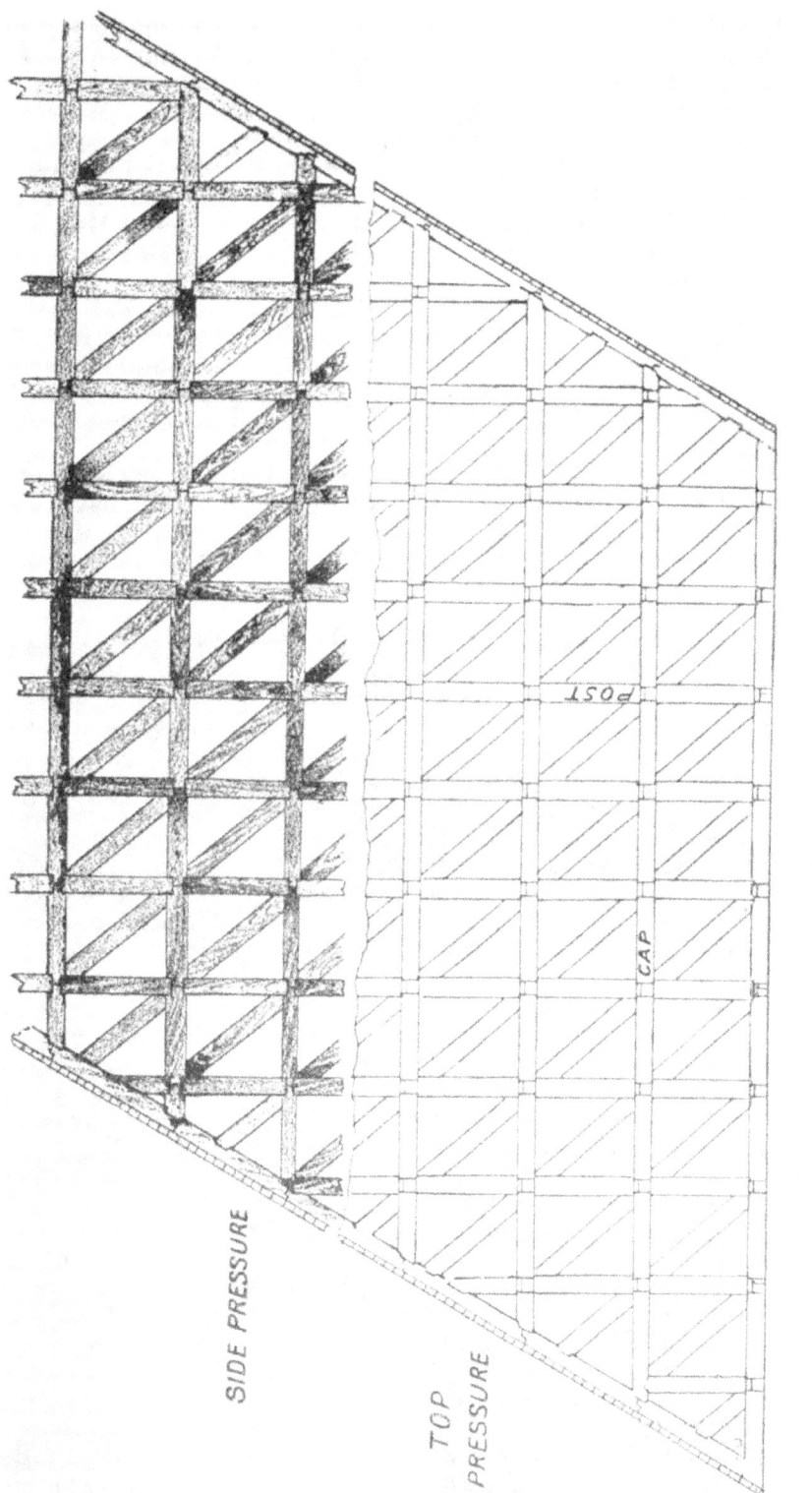

Fig. 61

posts in each case were set perpendicular. Returning to the point where these operations were begun, a second section by the side of the first cross-cut was taken out and timbered with a single line of posts and caps, the 2"x 4" scantling being placed as in the first case. When this section was completed there were standing three lines of posts surmounted by three lines of caps, extending from the foot to the hanging wall. This was not really a new idea, as Mr. Deidesheimer had previously employed the same method in his drift mine on Forest Hill, when the breast was carried in 125' wide, the roof being supported by rows of posts with continuous caps.

The work thus far performed in the Ophir revealed the fact that an extremely rich body of ore extended upward from the level where this work had been done. The miners were directed to commence stoping upward in the body of soft, black, crumbling ore, and soon a considerable excavation had been made. It became evident that the ground must be secured at once.

In the Georgetown Mine, which Mr. Deidesheimer had left but a short time before, the vein was vertical, and the walls were so soft and crumbling that, in order to stope out the mineral, he had resorted to the expedient of setting one post directly above another, the lower end resting on the cap, and in this way he managed to work the vein without much difficulty. (See Figs. 51 and 52.)

The Georgetown experience suggested the idea of adopting a similar plan in the Ophir. Accordingly, Mr. Deidesheimer had a mortise cut at the junction of two caps, which were already in place, and, having a post framed with a tenon to fit, set the post in place directly above the one resting on the floor below. In a short time four posts were in position with the caps upon them as below, together with the frail 2"x 4" scantling, the office of which was to keep the other timbers from falling down. The first "square set" timbers, it will be seen, were framed in the mine, the mortises being cut in the timbers in place. The work of extracting ore proceeded slowly yet, for the ground had to be secured as well as possible. It soon became evident, however, that something more substantial than 2"x 4" scantling would be required to keep the timbers in position, and it was determined to put in timbers of the same dimensions as those forming caps and posts. This was done at once, and the "square set" was complete in principle, though not in detail. The caps occupied all the space on top of the posts, leaving no resting place for the "ties," which had to be supported in some other manner. As they were looked upon as simply an auxiliary—a support to the posts and caps—they were only required to be held in position. Accordingly, a lot of iron spikes were made, in shape somewhat like the thumb, having a sharp point at one end, the other end having a face three fourths of an inch square. Two of these spikes were driven into a post at the proper height, and two in the post opposite, and the tie placed so that the ends rested upon these iron lugs, wedges being driven in at the ends to secure firmness. The posts and caps were now framed on the surface and delivered below, ready for use whenever needed. The work of mining now progressed much more rapidly and the problem seemed solved. Soon after it was determined to frame the timbers so that the ties might also rest on the posts. The stopes becoming of such great size, the dimensions of the timbers were increased, and they were then framed as shown in the accompanying illustrations. (See Figs. 60 and 61.)

As the work progressed slight changes were made from time to time whenever any improvement seemed possible. Sills were laid upon the floors of the levels as a foundation for the timbers above, which had now assumed massive proportions. The sill timbers were as long as it was possible to get into the mine. The men who were obliged to handle these ponderous timbers could see no reason why the sills should be longer than the caps, and had from the first looked upon the growth of this new system of timbering with much prejudice. When the great stopes were carried up from level to level the wisdom of the use of long sills became apparent, as they permitted the removal of all the ore and the placing of timbers without danger or loss, which could not have been accomplished with short sills, as, when breaking up through the floor of a level from below, short sills would have had nothing to sustain them.

When, in the course of ore extraction, the work reached the walls, additional timbers, called "wall plates," were put in, as shown in the sketch of timbering in the Ophir Mine. (See Fig. 60.) The caps were extended from the nearest post to the wall plate, except when a post came within 2' of the wall plate. In such case the cap extended from the wall plate to the second post in a single piece.

By a close inspection of the drawings the details may be plainly seen. It will be noticed that a system of braces reaching diagonally across the sets was also introduced, as well as close lagging on the walls. There is no doubt that the Ophir was the best timbered mine in the world; but the Comstock miners, who gladly adopted the Deidesheimer system, soon began to disregard many precautions which to them seemed unnecessary. The diagonal braces were left out, and the ground was found to stand about as well. Then, anxious to still further reduce expense and hurry the work of extraction, the lagging on the walls was dispensed with, and later, in somewhat firmer ground, the wall plates were left out, and finally the timbers were placed in rectangular sets, with only a few props here and there to the walls and roof, as shown in the drawings of the Caledonia Gold Mine. (See Figs. 62 and 63.) The disregard of these important details in American mines has resulted in numerous disastrous caves.

The sketch of the Caledonia Mine, in the Black Hills, South Dakota, represents the mine as it appeared in the spring of 1883. The main shaft was sunk in the large vein, the hoisting works being located on the adit level, 200' from the surface, vertical measurement, and 820' from the mouth of the tunnel. This shaft reached the foot wall on the 300-foot level, and a large stope was at once opened around the shaft, a pillar of ore 30' square being left to support it. The shaft was continued vertically to the 400-foot level, and an extensive stope opened there also. The gold-bearing rock of the Caledonia is hard white quartz, which occurs in bunches and reticulated veins in chloritic schist, and in this respect resembles some California mines, as the Utica-Stickle, Gover, and some others. In the Caledonia great headers were excavated in advance of placing the square sets. The timbers were all properly framed and were massive; but there was a disregard of what were considered the minor unnecessary details in placing them, particularly on the walls and against the roof. As these large stopes were extended, too broad an area was taken out at one time, and the superincumbent weight at length threw the timbers out of line, and almost without warning the mine caved, the immense timbers "jack-knifing" and snap-

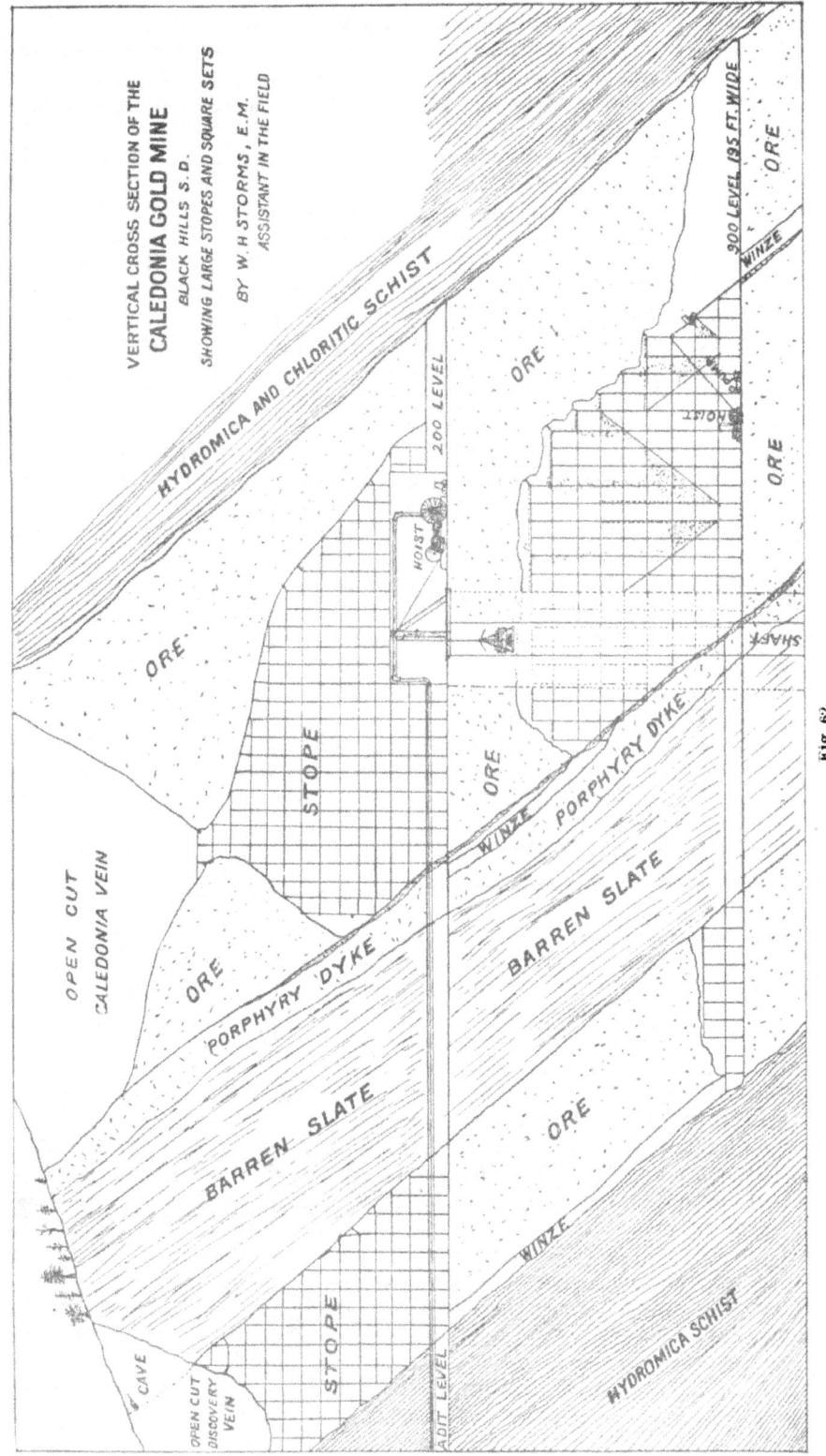

Fig. 62.

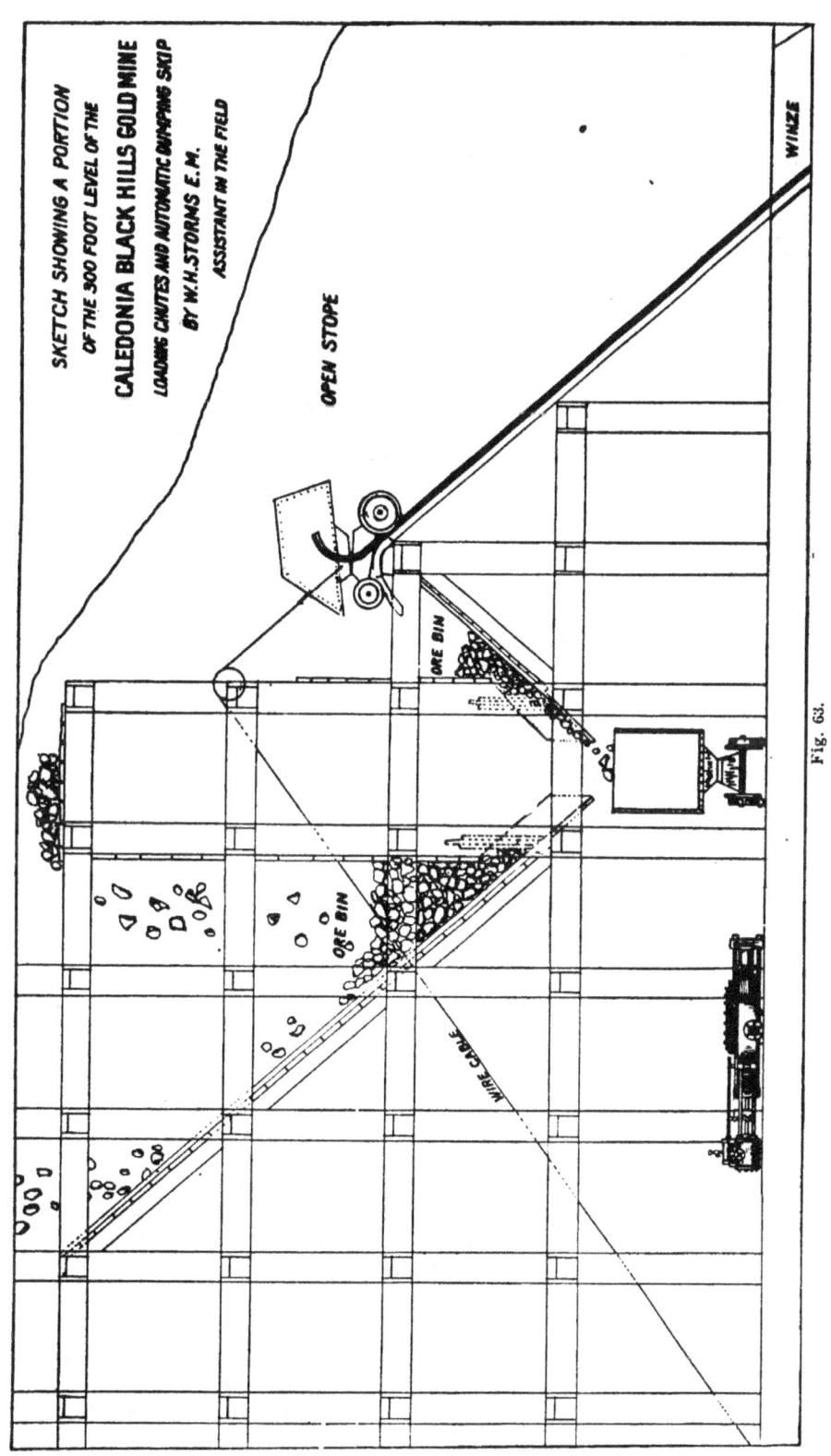

Fig. 63.

ping like reeds. Thousands of tons of ore, the expensive machinery, and a mass of splintered timbers were dumped in a chaotic mass to the bottom of the 400-foot level. There were three causes which led up to this disaster: The extraction of too large an area of ground at one time; carelessness in timbering, and the slippery nature of the foot wall (talc schist), which afforded a poor support to the large pillar of ore.

It is unreasonable to expect a frame of timbers, however strong, to support the weight of a mountain of rock, and for that reason discretion should be used in extracting large bodies of ore. It is hazardous to attempt to remove a section more than three or four sets wide at one time from the floor of one level to that next above. A breast may extend entirely across the deposit or vein, but if more than four sets in width are removed at one time it allows too much weight to fall upon the timbers, and the probability of a cave is greatly increased.

By taking out a section which may extend entirely across the vein or deposit, three or four sets wide, and carrying the stope up somewhat in the form of a pyramid, so that on nearing the floor next above only the space of a single set, or at most, not more than two, be at first removed, and the timbers firmly wedged to the sills before enlarging the excavation at that point, all the ore may be extracted, and the operation in this manner is attended with the least expense and danger. When a section has been mined out from level to level as described, a second section may be attempted. Under all ordinary circumstances it will be found that the timbers will usually support the ground until the stopes can be filled with waste, which should be done as fast as the ore is removed. At many mines waste is obtained on the surface, and in some the opening of large chambers in the hanging wall is necessary to obtain a sufficient amount of material for filling the stopes. There is always danger in the removal of a mass of rock which stands on a base that is broader than its upper portion, or apex, like the letter "A," while on the other hand a "V" shaped mass is largely supported by the walls. Many ore-bodies are lens-shaped; that is, broader at the center than those portions either above or below; and in these masses the greatest expense and danger attend the removal of the upper portion.

SIZE OF TIMBERS.

The size of timbers heretofore used in square-set timbering has ranged from 8" x 8", or 8" x 10", to 20" x 24", or even 24" x 28", though timbers of the latter dimensions are often difficult to obtain. Most California mines use very heavy timbers in large stopes, mostly round, and it is not uncommon to see them 24", and even 30", in diameter. With the use of these immense logs, filling has largely been neglected, to the detriment of the mines. There is a growing tendency to reduce the expense of timbering in large stopes by using timbers of smaller dimensions. Where formerly very heavy timbers were used in some of the largest mines of Butte, Montana, 12" x 12" timbers are now successfully employed, the stopes being filled with the waste.

CONSTRUCTION OF SQUARE SETS.

When a vein or ore-body is sufficiently large to require the square-set system of timbering, a stope is opened on an established level, and the construction of square sets is commenced by the placing of sills on the

METHODS OF MINE-TIMBERING.

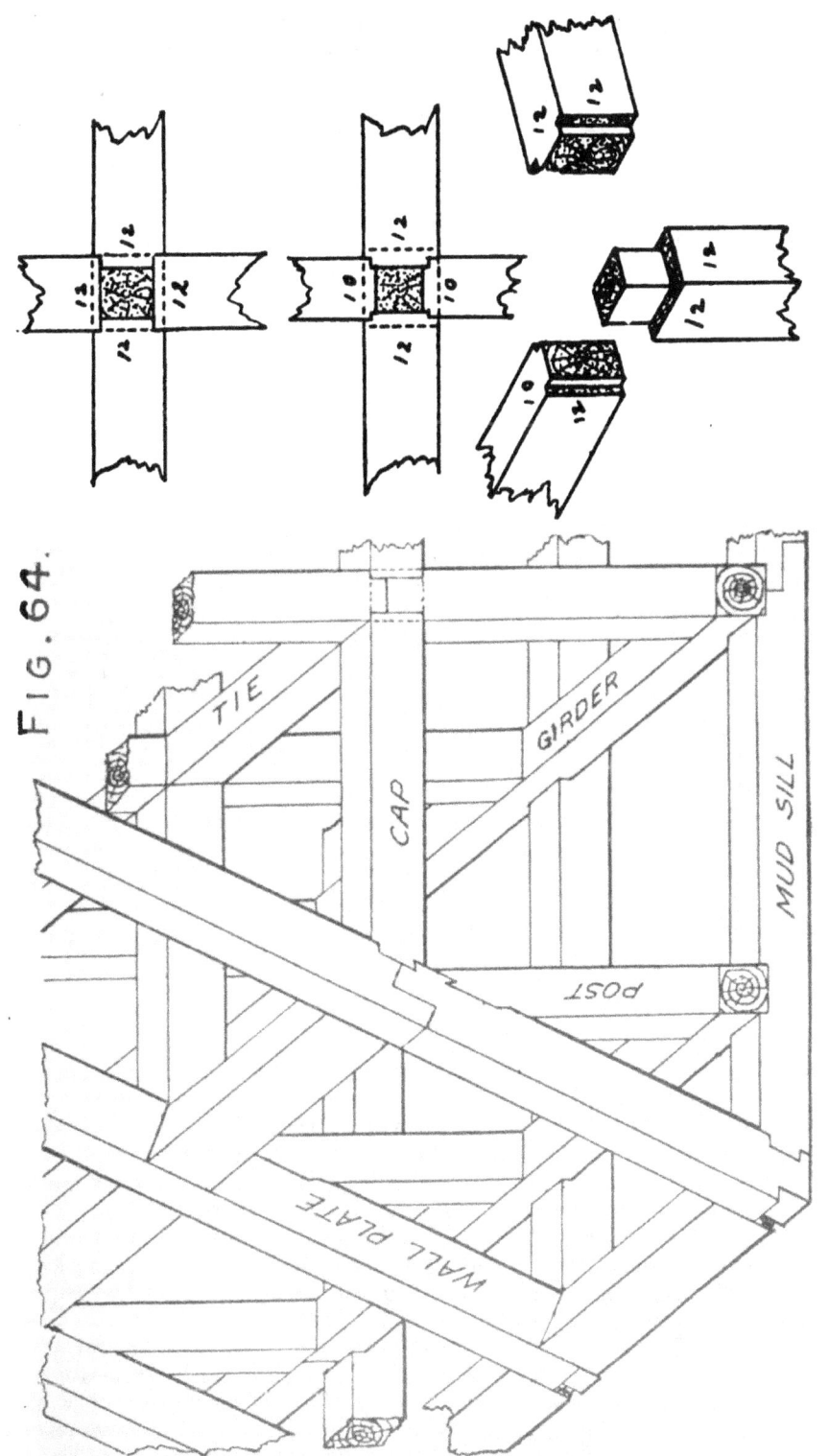

FIG. 64.

floor of the level. They are laid at right angles to the general strike of the vein or ore shoot, and must be exactly parallel. The sills should have a length equal to at least three full sets, 15', assuming that the five-foot system be adopted. If twenty-foot lengths can be obtained it is an advantage. These timbers are carefully framed to fit every intersection with other timbers. The end joining the wall plate is framed with a tenon, 1" long, extending entirely across the face of the sill. This fits into a mortise cut in the wall plate. (See Fig. 64.) At a certain distance (determined by measurement) from the wall plate, a "dap" or shallow depression, 12" x 12", is cut in the sill to the depth of 1", and beyond this point a similar dap at every 5' These are for the reception of the girders

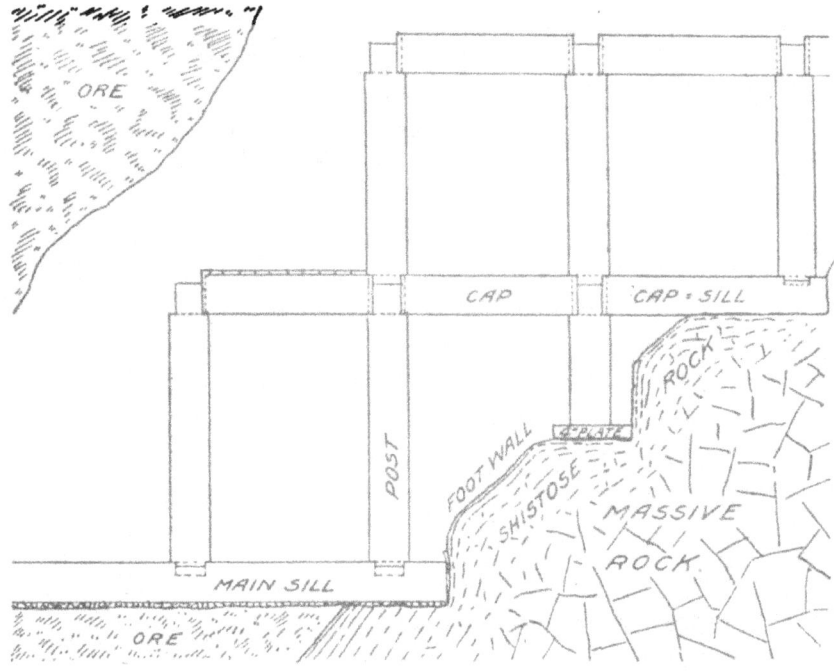

Fig. 65.

(sometimes called ties), which may be the length of at least two sets (10'), and are placed at right angles to the sills, extending longitudinally with the vein. (See Fig. 64.) When the ends of two sills meet, which should always be directly under a post, they must be spliced as shown in Fig. 64. The girders may also be spliced in a similar manner, or the ends may abut flush on the center of a sill. In the girders daps may also be cut 1" deep, with or without a mortise (8" x 8" and 1" deep) in the center for the insertion of a tenon on the lower end of the post; or, the dap may be dispensed with and the mortise only made. The object is to prevent the post shifting either from pressure or from rocks thrown by a blast. Fig. 64 shows the posts set without the dap in the girder; the mortise does not show. When short girders are used, as in many mines, the upper face of the sill may be framed as shown in Fig. 66. The girt or tie on either side forms with the sill a dap to receive the foot of the post, without a mortise. This method has some advantages, particularly the ease with which all rocks or other foreign substances may be removed when ready to set the post. The dap is cut 1" deep; a deeper one would afford no additional security or firmness.

Wall plates are framed as shown in Fig. 64. Where a joint occurs, which must always be opposite a post or cap, preferably the latter, the upper one should always overlap the lower one, except on the sill floor. (See Fig. 64.) It is not necessary that wall plates should exceed 6' or 8' in length. In some mines a continuous wall plate is dispensed with and short lengths of heavy plank are placed opposite each cap and post and wedged firmly, as shown in Figs. 68 and 69. Two styles of framing these timbers are shown in Fig. 64. Where posts and caps are each 12" x 12", girts (ties) 10" x 12" may be used, the 10" face being placed uppermost. In this case the post is framed with an 8" x 8" tenon on each end, 9" long on top and 3" long on the bottom. The cap has tenons 10" wide, and the girt tenons 8" wide, each projecting 1". But caps and girts have a 2" support on the post.

If 12" x 12" girts are used the cap is cut off square, or without a tenon, and is 50" long. Its rests its full width (1") on the post. The girt is also 50" long, and is so framed that 10" of its width will rest on the post, an inch on each side binding the cap.

Posts are framed with a tenon both top and bottom. Where 12" x 12" timbers have been adopted, these tenons are made 10" square; that on the upper end being 9" long, and that on the lower end 3" long. An exception to this may be made in the lower tenon of posts on the sill floor. Posts on the sill floor may also be longer than those on floors above to allow more freedom in moving about the gangways, loading chutes, etc., being on that floor. Those caps between the wall plate and the first post, which are less than 3' in length, are frequently cut in a single piece extending from the wall plate to the second post. In that case a dap 10" x 10" and 1" deep is cut at the proper place in the top, bottom, and sides of the cap. These will admit the girts on either side in the usual way, and with them form a shallow mortise on both the upper and lower sides of the cap for the reception of the posts, which are especially framed and provided with tenons 10" x 10" and 1" long. A glance at Fig. 60 makes it evident that many different combinations will occur along the line of the wall plates.

The sills and girders having been laid, the four posts of the first set are placed in position, and may be held in place temporarily by nailing strips of 1" x 6" to them. The caps are lifted to their places and additional posts are set up and caps added. The girts may then be slipped in from above and forced down until they rest upon the shoulders of the posts. The timbers are wedged firmly to the surrounding rock, either walls or ore. If the latter, they are, of course, only temporary.

In this manner the timbering progresses with the extraction of the ore. As the stopes are carried upward, set is built upon set until they reach from level to level. Tracks are laid on the sills, the floor being raised in the gangways to the level of the top of the sills by filling with rock. Girders may be left out of those sections where transverse gangways are necessary. As soon as practicable, bins are constructed in the sets above (see Fig. 63), provided with loading chutes. All ore broken above then falls or is shoveled into the bins, from which it is drawn into cars and sent to the surface. Temporary floors of lagging, 6" thick, are laid on the topmost sections of the square sets at certain points. These catch the ore as it is broken down, and the miners, standing on the broken rock, are brought within easy reach of the roof of the stope. These floors are moved from time to time as the work progresses.

When the timbers have been placed, the wedges along the walls should be tightly driven up between the wall and the plate, and in all other places where necessary to secure rigidness, and some one man should be detailed whose duty it is to look after them and see that they remain so; for in large stopes the ground is constantly shifting more or less and

Fig. 66.

the weight being transferred from one point to another causes some of the wedges to loosen, and they must be again tightened to insure safety.

DIAGONAL BRACES.

These timbers were at one time considered indispensable in heavy ground, but are seldom seen now. As cases may occur where it would seem advisable to use them, they will be briefly described. It must be borne in mind that they, in a measure, perform the function of stulls, and should not be placed at a right angle with the hanging wall, but at a higher angle, like stulls. For instance, suppose the inclination of the hanging wall be 45°, the diagonal braces should then stand at about 55°. It is then evident that the relative height and width of the sets must be determined by the angle it is desired to give the diagonal (angle) braces; before this can be determined, the general inclination of the hanging wall must be known. Angle braces may be the same size as posts, or if desired, somewhat smaller. They are slipped in sideways after other timbers of the set are in position, and are secured by shingle wedges, if not tight when put in place. (See Fig. 60.)

CAP SILLS, AND THEIR USE.

Cap sills are often used in preference to wall plates on the foot wall of a vein. They are applicable to small stopes as well as great chambers. As the name implies, these timbers are simply an extension of the cap from the post nearest the foot wall to the wall where it rests upon a hitch cut in the wall (2' or more, according to the character of the ground). It then forms a sill for the post of the set next above, as illustrated in the accompanying drawing. (See Fig. 66.)

SIDE PRESSURE.

When the pressure is greater, apparently, from the side than from overhead, the method of framing and joining these timbers is somewhat different. The cap becomes the post. An examination of the drawings (Fig. 67) will convey an idea of the system employed under these conditions. The tenons at each end of the cap in this case are exactly the same size, while the tie is also provided with a tenon, but of different style from that used where the timbers are framed to resist top pressure. The placing of wall plates and angle braces is the same in each system. The accompanying drawings, showing a horizontal and vertical section of a corner in both the top and side pressure systems, show plainly the details of both methods. (See Figs. 64 and 67.)

TIMBER CHUTES.

For the convenience of handling timbers in the mine, winzes should be cut in or near the foot wall of the vein, reaching from level to level. They are located at stated distances, 100' to 200' apart. Timbers to be used in the stope are lowered through the winze from the level above and delivered on the particular floor where they are to be used. This saves much unnecessary hoisting by hand underground. The winzes also improve ventilation.

IMPORTANCE OF FILLING.

Experience has proved that, by filling the stopes contemporaneously with the extraction of ore, mines are more safely and economically worked, with less loss of ore and fewer disastrous caves, than when heavier timbers were employed and the stopes left completely open. When additional strength is required immediately, the sets already in place are reinforced by placing additional posts on each side of those in the set, slipping them in between caps or girts so that each post of a set is surrounded; when this is insufficient, "bulkheads" of solid timbers, laid at right angles, or open cribs of timber filled with rock and reaching from the floor to the roof of the stope, are hastily constructed.

It would seem advisable to place upon the mudsills of each set, a floor of heavy lagging, 6" by 8", reaching from center to center of the sills. These may be placed 3" or 4" apart, or, if considered necessary, may be laid closely. A floor so provided will not only afford protection to miners in connecting levels, but also facilitate the recovery of all the ore without mixing it with the filling from the stope above. It may be pos-

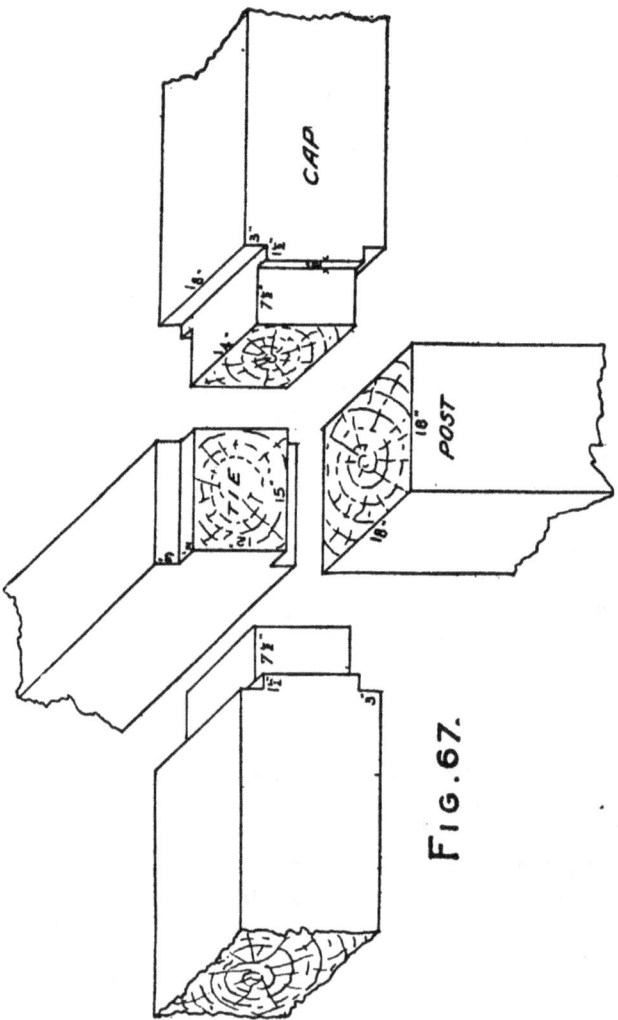

Fig. 67.

sible to reclaim a portion of this timber when connecting levels. The expense should not exceed 15 cents per ton of ore extracted from the sill-floor sets.

It has been repeatedly demonstrated in many mines that timber in any amount will not support great masses of rock which are settling with irresistible force. Experience has taught that stopes in large veins must be filled with waste rock, and the sooner this is done after the extraction of the ore the better. The great weight of rock on the hanging wall will compress timbers into a much smaller space, and a few inches in each set will amount to several feet in a large stope. It should be the aim to prevent, as far as possible, this shrinkage by filling. When it does occur great masses often become detached from the hanging wall. They have the appearance of being perfectly safe, and give no indication of danger when sounded with hammers; but wide, open spaces may exist back of them, and the heavy concussion of power drills or that produced by blasting, may cause the masses to fall with disastrous results.

RECOVERING LOST GROUND.

A great deal of good ore is sometimes lost by the settling of large veins upon the timbers of the first few floors of a stope so heavily as to throw the timbers out of line, when bulkheading and filling in must at once be resorted to, to prevent caving. In the Anaconda Mine, at Butte City, Montana, a case of this kind occurred. The timbers on the sill floor were forced out of line, and it was evident that a disastrous cave was imminent. The stope was packed full of timber and waste as rapidly as possible and the level abandoned. A drift was then run from a crosscut in the foot wall, 60' from the vein, and in this lateral drift, at distances of 60', and directly opposite the chutes in the abandoned stope, as established by the mine surveyor, a square set was placed. A raise was then carried up on an incline to the vein, where it was still intact above the filled stope from each of these points. These raises were timbered with square sets, which in section would present the appearance of a series of steps. The rock in which the drift was run was solid granite, but as the foot wall was approached it was much decomposed and required substantial timbering. Had the foot wall country rock been firm and hard no timbering would have been necessary, except at the stand on the floor of the drift where the loading chute was constructed and at the junction of the winze and foot wall of the vein. Through the winzes thus made the lost ore was almost entirely recovered.

BULKHEADS AND CRIBS.

It frequently occurs that bulkheads of solid timber, or cribs filled with rock, have to be built to sustain sinking roofs which have already been substantially timbered. Bulkheads are formed by laying massive squared timbers side by side, filling the space between sets. Upon these are laid a second layer at right angles to the first, and so on from floor to roof of the stope. Cribs are constructed by building up timbers in the form of a hollow square or rectangle, the interior space being filled with large broken rock, and also extend from floor to roof. Either of these devices forms a very substantial support for settling ground, though sometimes inefficient.

Bulkheads of solid masonry are sometimes built in mine workings to stop a heavy influx of water. They should be constructed with the greatest care and in the most substantial manner possible to prevent future collapse. When it becomes necessary to build a bulkhead of this kind a section should be cut in the floor, sides, and roof of the drift, and the masonry extended into these spaces. Water bulkheads are sometimes built of dry timbers, which, swelling, effectually shut off the influx of water.

An instance recently came to the knowledge of the writer in which a large flow of water followed the cutting of a subterranean reservoir by a power drill. As usual in such cases, wooden plugs were inserted and driven in, but to no purpose, the force of the water driving them out. The protection of a bulkhead seemed necessary to prevent the flooding of the mine. An experiment was first tried, which, proving successful, rendered the bulkhead unnecessary. A section of pipe provided with a globe valve was inserted in the hole, the valve remaining open until the pipe was firmly secured by wooden braces and tightly wedged in the rock. The valve was then closed and the flow shut off. Provision then being made to handle the water, the valve was opened partly and the reservoir drained. It seems that this plan might be adopted with advantage where the influx of water interferes with the construction and completion of a water-tight bulkhead, for at times the flow of water is so great that it is almost impossible to construct a bulkhead, owing to the pressure of the water back of it.

TIMBERING AT THE UTICA MINE.

The largest known body of ore in California to-day is in the Utica-Stickle Mine, at Angels Camp, Calaveras County. The workings have thus far exposed a mass of gold-bearing rock ranging from 40' to more than 100' broad, over 400' high, and exceeding 1,000' in length. Stopes of this size require a very substantial system of timbering, and the square-set system, which has been described at length, is exactly suited to its requirements. It is not in use, however, but in its stead another system, which is quite similar, and based on exactly the same principles, yet having very important differences. A study of the accompanying timber sketch of the Utica Mine will suffice to make an understanding of the method plain. (See Figs. 68 and 69.)

The timber used in this mine is exclusively round, peeled pine logs, which are delivered by contract at the surface works. The cost of these logs, I am informed, is 10c. for each inch of diameter, making a 16-foot 18" log cost $1 80, and a 24" log, $2 40. These timbers are all cut into two lengths of 8' each. They are mostly over 18" in diameter, and occasionally are found as large as 28" and 30" in diameter. The posts are framed to 14", top and bottom, the tenon being 4" long. Caps are framed to 14" at either end, the tenon being 6" long. Thus it will be seen when two adjacent caps are placed on a post the tenons will not meet, but will have a 2" space between them. This space is filled by driving in a section of 2" plank 14" square. At the junction of caps and posts, it will be noticed, are two ties (sprags) in place of one, as in the square-set system. These ties range from 12" to 16" in diameter, and are about 4' in length. The lower one has a horn 4" in length on its upper half. These projections rest upon the shoulders of the post. The upper tie is tightly wedged between the posts, as shown in the drawing. This system has replaced that formerly in use at the Utica Mine, in which the posts were 16' high; caps being let into the sides at 8' and others resting on the top. The new system is much preferred and has many advantages over the old (a 30" post 16' in length weighed over 3,300 pounds). In the present system the posts and caps weigh from 700 to over 1,000 pounds each.

Owing to the impossibility of procuring round timbers that are exactly uniform in size, the Utica system involves a vast amount of dressing timbers underground to secure the necessary uniformity in joining. While the system is one affording great strength when properly placed,

Fig. 68.

it would seem to possess no advantages over the square-set system, and to be more troublesome, cumbersome, and, in the writer's opinion, more expensive than the latter, though the first cost of round timber is less than that of sawed square timber.

Sills have only recently been introduced in the Utica system. The wall plates consist of short, single pieces of heavy plank, which are placed between the posts or caps and the rock-mass. (See Figs. 68 and 69.)

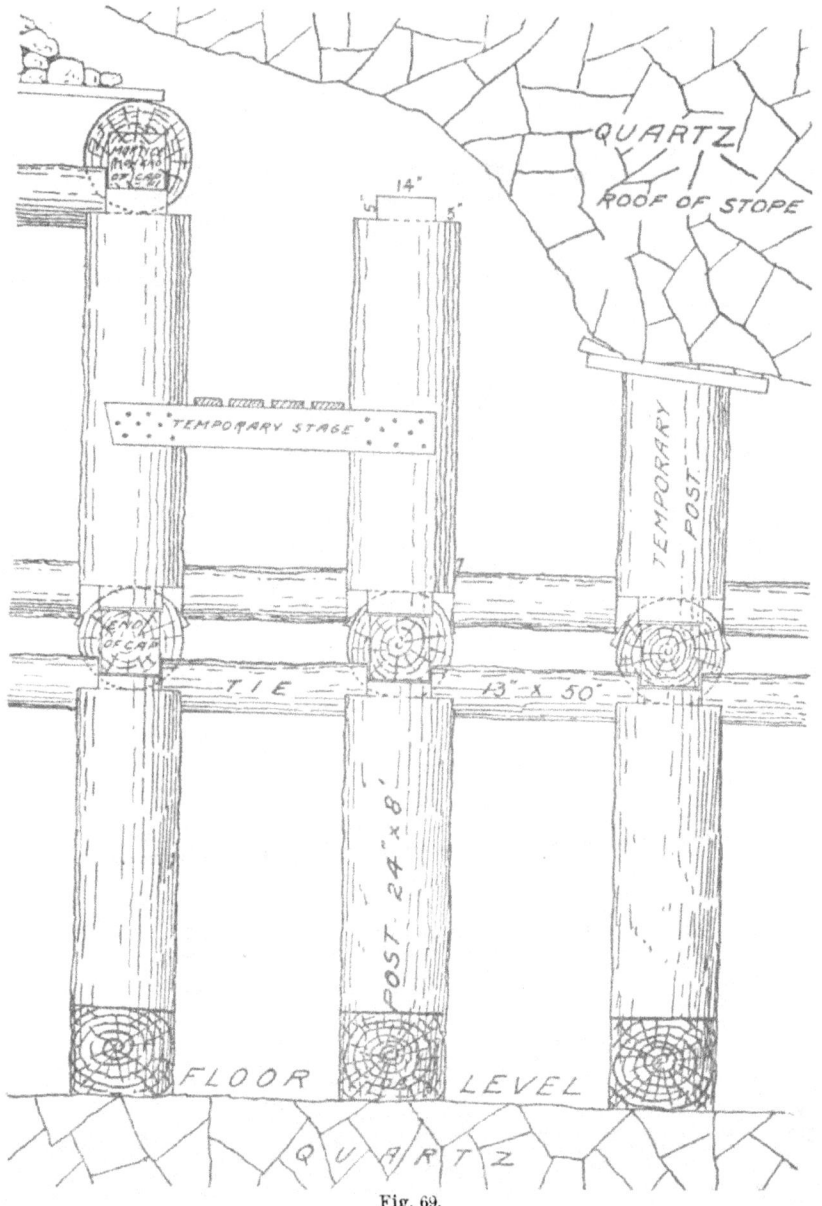

Fig. 69.

SYSTEM AT THE WILDMAN MINE.

The system of timbering at the Wildman Mine at Sutter Creek, Amador County, California, is quite similar to that above described. The posts are provided with tenons 6"x 12", which on top of the post are $8\frac{1}{2}$" long, and on the bottom $3\frac{1}{2}$" long. The posts rest upon each other, the caps being framed with a tenon 12"x 12", which is 4" long. These tenons rest upon the shoulders of the posts. Double sprags are also employed in this system, which are so placed as to abut against both the upper and lower posts at their intersection. The right-hand sprag binds the cap on the right, and the left-hand sprag the cap on that side. This method, like that of the Utica, is simply a modification of the square-set system. The Wildman timbers are round, and usually about 14" in diameter.

HANDLING TIMBERS.

The manner of handling timbers at mines depends to a great extent on the size and number of timbers used, and on the character of the shaft, whether it be vertical or inclined. In drift mines and mineral deposits which lie nearly horizontal, they are sent in on flat cars (trolleys) built for the purpose. When cages are used in vertical shafts the timbers are set upright on the floor of the cage, lashed together, and lowered into the mine. At inclines it is a common practice to load the timbers into the skip and in that way lower them down the shaft to the desired point.

At the Utica Mine the heavy timbers are handled quickly and easily on the surface. They are sent from the framing machine to the shafts on trolleys. Two men are kept busy a great deal of the time in sending timbers underground. A number of stout chains 8' in length are provided. In the center of each chain is a 4" ring, and at each end a dog, having a point 4" in length projecting at about 70°. One of these dogs is driven into the side of a post or cap somewhat above the center. The chain is passed over the end of the timber and the other dog is driven into the opposite side. A spike, 6" to 8" long, having a 5' piece of rope attached, is driven well into the lower end of the timber, and it is ready to lower into the mine. Beneath the skips are stout chains with rings. An extra chain having hooks at the ends is at hand. This forms the connecting link between the chain beneath the skip and that attached to the timber. Having been securely hooked the skip is raised slowly, lifting the timber from the platform. The rope and spike at the bottom are used to bring the timber to a standstill before the skip is lowered. Having reached the level where it is wanted the rope again comes into use in landing the timber at the station.

The Utica shaft is almost vertical. This process, while a convenient and quick way of handling these heavy timbers, cannot well be used in shafts departing very far from the perpendicular, unless a slide of plank be laid between the guides or runway of the skip, and it would then be advisable to line the slides with strips of flat iron. It would appear that timbers might be handled in this manner in shafts having as low an inclination as 35°, particularly if the slide be lubricated. It is certainly a superior method to that of loading the timbers into skips, which involves much difficulty, particularly in landing them at stations. Most mines have blocks and tackle at the stations, secured to heavy timbers overhead, for unloading timbers from skips.

In mines where cages are used, when it is necessary to send long timbers down into the mine, as sills for square sets, or long plates, much time is saved by boring an auger hole through the timber near the end and passing an iron bar ($\frac{7}{8}$") through it. The ends of the bar are provided with threads and nuts. These ends pass through the ends of a

U-shaped frame, forming a large clevis. The clevis is attached by a chain to the bottom of the cage, and in this manner lowered to the point desired. Fig. 70 will convey an idea of the device above described.

All timbers should be framed on the surface where possible, as it saves much time and trouble. Gang saws are much to be preferred to any other method of framing timbers for square sets. The timbers can, in

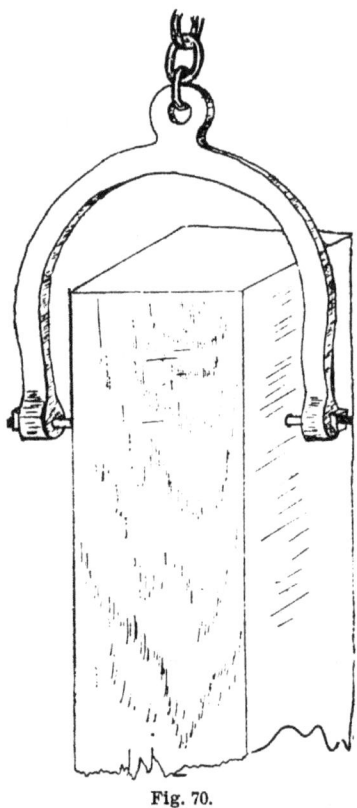

Fig. 70.

this way, be framed as exactly if not more so than by hand, and in one tenth of the time.

TIMBER TROLLEYS.

At some mines where the working-shafts are inclines, timber cars or "trolleys" are in use; they are lowered into the mine beneath the skip to which they are attached, the wheels running on the same track as the skip. Skids are arranged at the station on each level and on the surface in such a manner that the trolley may be drawn across the shaft and onto the main trackway. When timbers are about to be sent down into the mine they are securely lashed to the trolley. The skip is drawn up above the level of the platform, the skids thrown across the shaft, and the trolley chain attached to the bottom of the skip. Both skip and trolley are then hoisted until the trolley clears the skids, which are removed and the timbers lowered in this manner to the desired level, where the skids on that level are put in place and the trolley is landed on the floor of the station, from which point the timbers may be sent to that portion of the level where they are required, without unloading from the trolley.

INDEX.

A

	PAGE.
Alma shaft, Jackson, Cal.	24-27-34
Anaconda Mine, Montana	61
Angle braces, in square sets	58
Angles, taking in shafts	34
Argonaut incline, Jackson, Cal.	33-34

B

Bark on timbers	9
Bedrock, swelling	14-15
Beveled notch, the	9
Bimetallic Mine	41
Blanket veins, stopes in	36
"Blocky ground"	19
Bolts and dogs for hanging shaft timbers	22
Bowlders from roof	19
Braces, diagonal	58
Breast-boards, method of using	18-30
Breasting caps	7-9
Buchanan Mine, Tuolumne County	44
Bulkheads and cribs for support of roofs	61-62
Bulkheads, water-tight	61-62

C

Caissons, where used	30
Calaveras Consolidated tunnel, timber in	7
Caledonia Mine, timbering of	51-52-53
California mines, square sets in	46
Cap sills and their use	59
Caps and posts, in drifts	9
In square sets	56
Size of	13
Careless timbering in Caledonia Mine	51-54
Caved workings, timbering	17-18
Chambers and stopes	34
Chambers, great	46-47
Chutes or ore-slides	40
Chutes, loading, in drifts	40-45-46
In shafts	40
Clay linings for shafts	30
Comstock shafts	24-25-26
Connecting levels	41-42
Construction of square sets	54-55-56-57
Cottonwood for timbers	6
Cribbed shaft	21
Cribs and bulkheads	61-62

D

Deidesheimer's square sets	47-48-49-50
Diagonal braces	48-58-59
Dividers in shafts	23-24-25-26-30
Dogs for hanging timbers	22
Drainage, importance of	13
Drains	11
Drift mines, timbering in	7-9-15-16
Drift of Kennedy Mine, timber in	10
Drifts and tunnels	7-8-9-10-11

E

El Capitan shaft, Nevada County, Cal.	28

INDEX.

F

	Page.
False sets	10–12
Filling, importance of	59
Fissures, precautions with	19–20
Flat veins, stopes in	36
Floors, temporary	57
Foot-blocks on swelling ground	17–18
"Fore-poling"	28
Forman shaft, Comstock Lode	24–25
Framing of shaft timbers	20–21–25–26
Of drift and tunnel timbers	9
Of guides in shafts	31–32
Of square-set timbers	55–56–57–58–59
Freezing quicksand and watery ground	30

G

Golden Gate Mine, Sonora, loading chute in shaft of	46
Gover Mine, timber tests in	6
Great chambers and square sets	38–39
Guides in shafts	31

H

Hanging bolts and dogs for shafts	22–23
Hanging bolts and dogs, manner of using	22–23
Hanging wall, timbering a soft	35
Handling timbers, on the surface at Utica Mine	62
Handling timbers, underground	62
Honduras Mine, timbering in	6

I

Inclines	32–34
Timbering	32–34
Argonaut, at Jackson, Cal.	34
Iron dogs for hanging timbers	22

K

Kinds of timber used in mines	6
Kennedy Mine, timbering in	10

L

Lagging, how made	10
In drift mines	7
In shafts	28
Kinds of wood used in making	12
Manner of driving in drifts	10
Manner of driving in shafts	10–28
Size of shafts	10
Large ore-bodies, stoping	41
Levels, connecting	41–42
Loading-chutes, manner of constructing	40–41–45–46
In drifts	40–45–46
In shafts	40
Square sets	46
Long wall system	36
Lost ground, recovering	61
Lowering timbers	65–66

M

Manner of stoping large ore-bodies	41
Measuring distance between walls	34–35
Montana method of placing dividers	30

N

New Almaden Quicksilver Mine, timbering in	35

O

Oak for timbers	6
Ophir Mine	47–48–50–51
Ore-bins	46
Ore-handling on underhand stoping	40–41
Ore shoot	40
Ore-slides or chutes	40–41–45–46
Overhand stoping	37

INDEX.

P

	PAGE.
Pillar and stall system	36
Pillars of ore to support roof	36
Plank under caps	9
Post and sill joint	13
Post in square set	55-56-57
Post in drift	7-9-15
Post and drifting cap	9
Principles of timbering mines	5

Q

Quicksand, freezing of	30

R

Railroad ties for timbers	21
Reachers in shafts	22
Redwood for timbers	6
Recovering lost ground	61
Requa shaft, Comstock Lode	23-26
Retimbering shafts	23
Rowland's method of timbering	17-18
Round timbers, to prevent decay of	9
Round shafts, where seen	30-31
Running ground, timbering in	17-18-19
To take advantage of	17

S

Saddle-backs	43-44-45
Saddle-wedges	44
Seat for post on sill	58
Serpentine, timbering in	19
Shafts, Alma, at Jackson, Cal.	24-27-34
Argonaut, Jackson, Cal. (inclined)	33-34
Cribbed	21
Comstock	23-24-25-26
Dividers in	23-24
El Capitan	28
Framing timbers for	20-21-25-26
Forman	24-25
Hanging timbers in	22-23
Inclined	24
In running ground	28
"L"-shaped	24-31
Linings in	21-25-26
Methods of timbering	5
Prospecting	19-20
Reachers in	22
Retimbering	23
Round	30-31
Size and shape of	19-20-24
Single compartment	21-23
Sinking in sections	23
Without timber	20
With two or more compartments	23-24-25-26
Side pressure	5-49-59
Sills for soft ground in drifts	7
Sill and post joint	13
For square sets	51
In swelling ground	17
Single sets in drift mines	15-16
Slips, danger near	17-19
Precautions with	19
Soft walls, timbering in	35-39
Spiked plank in drifts	9
Spliced wall-plates	28
Sprags	12
Spruce for timbers	6
Square sets, Deidesheimer's system	47-48-49-50-51
Building up	57
Diagonal braces in	58-59
Floors in	57
Framing timbers for	50-51-54-55-56-57-58
Girders in	55-56
In California mines	46-47
In South Dakota	51-52-53

	PAGE.
Square sets, invention of	47-48-50
Method of constructing	54-55-56-57
On Comstock	47-48-49-50
Ore-bins in	46
Post, cap, tie, sill, wall-plates	55-56-57
Side pressure on	49-58
Sills in	54-55
Size of timbers of	54
Strength of	56-57
Timbers of	56-58
Top pressure on	49
Wedges, importance of	58
Stations, timbers in	31
Stopes in flat veins	36
And chambers	34
Bulkheads in	61-62
In steeply inclined veins	37
Loading-chutes in	45-46
Ore-bins in	45-46
Recovery of timber from	36
Stoping overhead	37-39
Large ore-bodies	40
Underhand	41
Strength of timbers	44-56-57
Stulls	5
And wedges	37
Distance between	37-39
In desert regions	20
In soft walls	37
Proper length of	37
Strength of	44
Sugar pine timbers	6
Sutro Tunnel	11
Swelling ground, danger of	5-14-15-17
In Hidden Treasure Mine	14
In Hardenburg Mine	14
In Sutro Tunnel	11
Method of timbering in	15
Peculiar system of timber in mines on Sugar Loaf Mountain	14-15
Use of breast-boards in	30

T

Talc ground, timbering	19
Tanks	46
Ties in drifts	13
In square sets	55-56-57
Timbers, falling out	41
Kinds used	6
Lowering	22-23-65-66
Round	9-62-63-64
Size of	54
Strength of	44
Timber chutes	59
Timber trolleys	66
Timberless regions, mining in	20
Timbering in soft walls	37-38-39
Track-laying	8-13
Trolleys for timber	66
Tunnels and drifts	7
Two-compartment shafts	23

U

Underhand stoping	40-41
Utica Mine, timbering	62-63-64-65-66

V

Veins, blanket and flat	36
Large in California	46-47
Steeply inclined	37
Stopes in	34-37-40
Soft walls in	35-37
Vicinity of	19

INDEX.

W

	PAGE.
Wall-plates, spliced	28-30-57
Walls, measuring distance between	34-35
Soft	37-39-40
Wedges, importance of watching	20-58
Saddle	44
Wildman Mine, retimbering shaft of	23
Wildman Mine, system	64-65
Winzes, for handling ore	40
For ventilation	40

Y

Yellow pine timbers	6
Yucca wood timbers	6

Z

Zeila Mine, California	47

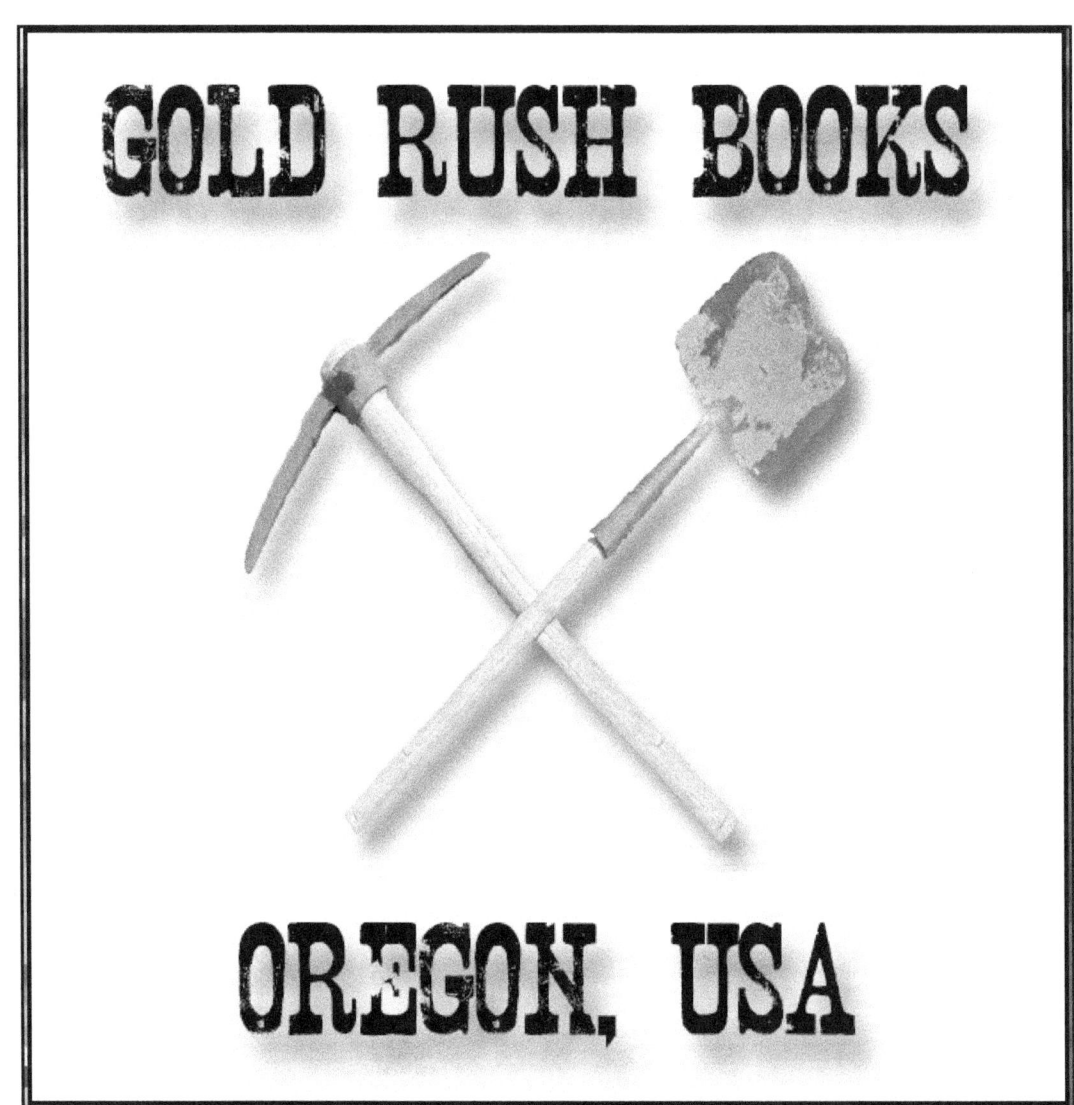

www.GoldMiningBooks.com

More Books On Mining

Visit: www.goldminingbooks.com to order your copies or ask your favorite book seller to offer them.

Mining Books by Kerby Jackson

Gold Dust: Stories From Oregon's Mining Years

Oregon mining historian and prospector, Kerby Jackson, brings you a treasure trove of seventeen stories on Southern Oregon's rich history of gold prospecting, the prospectors and their discoveries, and the breathtaking areas they settled in and made homes. 5" X 8", 98 ppgs. Retail Price: $11.99

The Golden Trail: More Stories From Oregon's Mining Years

In his follow-up to "Gold Dust: Stories of Oregon's Mining Years", this time around, Jackson brings us twelve tales from Oregon's Gold Rush, including the story about the first gold strike on Canyon Creek in Grant County, about the old timers who found gold by the pail full at the Victor Mine near Galice, how Iradel Bray discovered a rich ledge of gold on the Coquille River during the height of the Rogue River War, a tale of two elderly miners on the hunt for a lost mine in the Cascade Mountains, details about the discovery of the famous Armstrong Nugget and others. 5" X 8", 70 ppgs. Retail Price: $10.99

Oregon Mining Books

Geology and Mineral Resources of Josephine County, Oregon

Unavailable since the 1970's, this important publication was originally compiled by the Oregon Department of Geology and Mineral Industries and includes important details on the economic geology and mineral resources of this important mining area in South Western Oregon. Included are notes on the history, geology and development of important mines, as well as insights into the mining of gold, copper, nickel, limestone, chromium and other minerals found in large quantities in Josephine County, Oregon. 8.5" X 11", 54 ppgs. Retail Price: $9.99

Mines and Prospects of the Mount Reuben Mining District

Unavailable since 1947, this important publication was originally compiled by geologist Elton Youngberg of the Oregon Department of Geology and Mineral Industries and includes detailed descriptions, histories and the geology of the Mount Reuben Mining District in Josephine County, Oregon. Included are notes on the history, geology, development and assay statistics, as well as underground maps of all the major mines and prospects in the vicinity of this much neglected mining district. 8.5" X 11", 48 ppgs. Retail Price: $9.99

The Granite Mining District
Notes on the history, geology and development of important mines in the well known Granite Mining District which is located in Grant County, Oregon. Some of the mines discussed include the Ajax, Blue Ribbon, Buffalo, Continental, Cougar-Independence, Magnolia, New York, Standard and the Tillicum. Also included are many rare maps pertaining to the mines in the area. 8.5" X 11", 48 ppgs. Retail Price: $9.99

Ore Deposits of the Takilma and Waldo Mining Districts of Josephine County, Oregon
The Waldo and Takilma mining districts are most notable for the fact that the earliest large scale mining of placer gold and copper in Oregon took place in these two areas. Included are details about some of the earliest large gold mines in the state such as the Llano de Oro, High Gravel, Cameron, Platerica, Deep Gravel and others, as well as copper mines such as the famous Queen of Bronze mine, the Waldo, Lily and Cowboy mines. This volume also includes six maps and 20 original illustrations. 8.5" X 11", 74 ppgs. Retail Price: $9.99

Metal Mines of Douglas, Coos and Curry Counties, Oregon
Oregon mining historian Kerby Jackson introduces us to a classic work on Oregon's mining history in this important re-issue of Bulletin 14C Volume 1, otherwise known as the Douglas, Coos & Curry Counties, Oregon Metal Mines Handbook. Unavailable since 1940, this important publication was originally compiled by the Oregon Department of Geology and Mineral Industries includes detailed descriptions, histories and the geology of over 250 metallic mineral mines and prospects in this rugged area of South West Oregon. 8.5" X 11", 158 ppgs. Retail Price: $19.99

Metal Mines of Jackson County, Oregon
Unavailable since 1943, this important publication was originally compiled by the Oregon Department of Geology and Mineral Industries includes detailed descriptions, histories and the geology of over 450 metallic mineral mines and prospects in Jackson County, Oregon. Included are such famous gold mining areas as Gold Hill, Jacksonville, Sterling and the Upper Applegate. 8.5" X 11", 220 ppgs. Retail Price: $24.99

Metal Mines of Josephine County, Oregon
Oregon mining historian Kerby Jackson introduces us to a classic work on Oregon's mining history in this important re-issue of Bulletin 14C, otherwise known as the Josephine County, Oregon Metal Mines Handbook. Unavailable since 1952, this important publication was originally compiled by the Oregon Department of Geology and Mineral Industries includes detailed descriptions, histories and the geology of over 500 metallic mineral mines and prospects in Josephine County, Oregon. 8.5" X 11", 250 ppgs. Retail Price: $24.99

Metal Mines of North East Oregon
Oregon mining historian Kerby Jackson introduces us to a classic work on Oregon's mining history in this important re-issue of Bulletin 14A and 14B, otherwise known as the North East Oregon Metal Mines Handbook. Unavailable since 1941, this important publication was originally compiled by the Oregon Department of Geology and Mineral Industries and includes detailed descriptions, histories and the geology of over 750 metallic mineral mines and prospects in North Eastern Oregon. 8.5" X 11", 310 ppgs. Retail Price: $29.99

Metal Mines of North West Oregon

Oregon mining historian Kerby Jackson introduces us to a classic work on Oregon's mining history in this important re-issue of Bulletin 14D, otherwise known as the North West Oregon Metal Mines Handbook. Unavailable since 1951, this important publication was originally compiled by the Oregon Department of Geology and Mineral Industries and includes detailed descriptions, histories and the geology of over 250 metallic mineral mines and prospects in North Western Oregon. 8.5" X 11", 182 ppgs. Retail Price: $19.99

Mines and Prospects of Oregon

Mining historian Kerby Jackson introduces us to a classic mining work by the Oregon Bureau of Mines in this important re-issue of The Handbook of Mines and Prospects of Oregon. Unavailable since 1916, this publication includes important insights into hundreds of gold, silver, copper, coal, limestone and other mines that operated in the State of Oregon around the turn of the 19th Century. Included are not only geological details on early mines throughout Oregon, but also insights into their history, production, locations and in some cases, also included are rare maps of their underground workings. 8.5" X 11", 314 ppgs. Retail Price: $24.99

Lode Gold of the Klamath Mountains of Northern California and South West Oregon

(See California Mining Books)

Mineral Resources of South West Oregon

Unavailable since 1914, this publication includes important insights into dozens of mines that once operated in South West Oregon, including the famous gold fields of Josephine and Jackson Counties, as well as the Coal Mines of Coos County. Included are not only geological details on early mines throughout South West Oregon, but also insights into their history, production and locations. 8.5" X 11", 154 ppgs. Retail Price: $11.99

Chromite Mining in The Klamath Mountains of California and Oregon

(See California Mining Books)

Southern Oregon Mineral Wealth

Unavailable since 1904, this rare publication provides a unique snapshot into the mines that were operating in the area at the time. Included are not only geological details on early mines throughout South West Oregon, but also insights into their history, production and locations. Some of the mining areas include Grave Creek, Greenback, Wolf Creek, Jump Off Joe Creek, Granite Hill, Galice, Mount Reuben, Gold Hill, Galls Creek, Kane Creek, Sardine Creek, Birdseye Creek, Evans Creek, Foots Creek, Jacksonville, Ashland, the Applegate River, Waldo, Kerby and the Illinois River, Althouse and Sucker Creek, as well as insights into local copper mining and other topics. 8.5" X 11", 64 ppgs. Retail Price: $8.99

Geology and Ore Deposits of the Takilma and Waldo Mining Districts

Unavailable since the 1933, this publication was originally compiled by the United States Geological Survey and includes details on gold and copper mining in the Takilma and Waldo Districts of Josephine County, Oregon. The Waldo and Takilma mining districts are most notable for the fact that the earliest large scale mining of placer gold and copper in Oregon took place in these two areas. Included in this report are details about some of the earliest large gold mines in the state such as the Llano de Oro, High Gravel, Cameron, Platerica, Deep Gravel and others, as well as copper mines such as the famous Queen of Bronze mine, the Waldo, Lily and Cowboy mines. In addition to geological examinations, insights are also provided into the production, day to day operations and early histories of these mines, as well as calculations of known mineral reserves in the area. This volume also includes six maps and 20 original illustrations. 8.5" X 11", 74 ppgs. Retail Price: $9.99

Gold Mines of Oregon

Oregon mining historian Kerby Jackson introduces us to a classic work on Oregon's mining history in this important re-issue of Bulletin 61, otherwise known as "Gold and Silver In Oregon". Unavailable since 1968, this important publication was originally compiled by geologists Howard C. Brooks and Len Ramp of the Oregon Department of Geology and Mineral Industries and includes detailed descriptions, histories and the geology of over 450 gold mines Oregon. Included are notes on the history, geology and gold production statistics of all the major mining areas in Oregon including the Klamath Mountains, the Blue Mountains and the North Cascades. While gold is where you find it, as every miner knows, the path to success is to prospect for gold where it was previously found. 8.5" X 11", 344 ppgs. Retail Price: $24.99

Mines and Mineral Resources of Curry County Oregon

Originally published in 1916, this important publication on Oregon Mining has not been available for nearly a century. Included are rare insights into the history, production and locations of dozens of gold mines in Curry County, Oregon, as well as detailed information on important Oregon mining districts in that area such as those at Agness, Bald Face Creek, Mule Creek, Boulder Creek, China Diggings, Collier Creek, Elk River, Gold Beach, Rock Creek, Sixes River and elsewhere. Particular attention is especially paid to the famous beach gold deposits of this portion of the Oregon Coast. 8.5" X 11", 140 ppgs. Retail Price: $11.99

Chromite Mining in South West Oregon

Originally published in 1961, this important publication on Oregon Mining has not been available for nearly a century. Included are rare insights into the history, production and locations of nearly 300 chromite mines in South Western Oregon. 8.5" X 11", 184 ppgs. Retail Price: $14.99

Mineral Resources of Douglas County Oregon

Originally published in 1972, this important publication on Oregon Mining has not been available for nearly forty years. Included are rare insights into the geology, history, production and locations of numerous gold mines and other mining properties in Douglas County, Oregon. 8.5" X 11", 124 ppgs. Retail Price: $11.99

Mineral Resources of Coos County Oregon

Originally published in 1972, this important publication on Oregon Mining has not been available for nearly forty years. Included are rare insights into the geology, history, production and locations of numerous gold mines and other mining properties in Coos County, Oregon. 8.5" X 11", 100 ppgs. Retail Price: $11.99

Mineral Resources of Lane County Oregon

Originally published in 1938, this important publication on Oregon Mining has not been available for nearly seventy five years. Included are extremely rare insights into the geology and mines of Lane County, Oregon, in particular in the Bohemia, Blue River, Oakridge, Black Butte and Winberry Mining Districts. 8.5" X 11", 82 ppgs. Retail Price: $9.99

Mineral Resources of the Upper Chetco River of Oregon: Including the Kalmiopsis Wilderness

Originally published in 1975, this important publication on Oregon Mining has not been available for nearly forty years. Withdrawn under the 1872 Mining Act since 1984, real insight into the minerals resources and mines of the Upper Chetco River has long been unavailable due to the remoteness of the area. Despite this, the decades of battle between property owners and environmental extremists over the last private mining inholding in the area has continued to pique the interest of those interested in mining and other forms of natural resource use. Gold mining began in the area in the 1850's and has a rich history in this geographic area, even if the facts surrounding it are little known. Included are twenty two rare photographs, as well as insights into the Becca and Morning Mine, the Emmly Mine (also known as Emily Camp), the Frazier Mine, the Golden Dream or Higgins Mine, Hustis Mine, Peck Mine and others. 8.5" X 11", 64 ppgs. Retail Price: $8.99

Gold Dredging in Oregon

Originally published in 1939, this important publication on Oregon Mining has not been available for nearly seventy five years. Included are extremely rare insights into the history and day to day operations of the dragline and bucketline gold dredges that once worked the placer gold fields of South West and North East Oregon in decades gone by. Also included are details into the areas that were worked by gold dredges in Josephine, Jackson, Baker and Grant counties, as well as the economic factors that impacted this mining method. This volume also offers a unique look into the values of river bottom land in relation to both farming and mining, in how farm lands were mined, re-soiled and reclamated after the dredges worked them. Featured are hard to find maps of the gold dredge fields, as well as rare photographs from a bygone era. 8.5" X 11", 86 ppgs. Retail Price: $8.99

Quick Silver Mining in Oregon

Originally published in 1963, this important publication on Oregon Mining has not been available for over fifty years. This publication includes details into the history and production of Elemental Mercury or Quicksilver in the State of Oregon. 8.5" X 11", 238 ppgs. Retail Price: $15.99

Mines of the Greenhorn Mining District of Grant County Oregon

Originally published in 1948, this important publication on Oregon Mining has not been available for over sixty five years. In this publication are rare insights into the mines of the famous Greenhorn Mining District of Grant County, Oregon, especially the famous Morning Mine. Also included are details on the Tempest, Tiger, Bi-Metallic, Windsor, Psyche, Big Johnny, Snow Creek, Banzette and Paramount Mines, as well as prospects in the vicinities in the famous mining areas of Mormon Basin, Vinegar Basin and Desolation Creek. Included are hard to find mine maps and dozens of rare photographs from the bygone era of Grant County's rich mining history. 8.5" X 11", 72 ppgs. Retail Price: $9.99

Geology of the Wallowa Mountains of Oregon: Part I (Volume 1)

Originally published in 1938, this important publication on Oregon Mining has not been available for nearly seventy five years. Included are details on the geology of this unique portion of North Eastern Oregon. This is the first part of a two book series on the area. Accompanying the text are rare photographs and historic maps. 8.5" X 11", 92 ppgs. Retail Price: $9.99

Geology of the Wallowa Mountains of Oregon: Part II (Volume 2)

Originally published in 1938, this important publication on Oregon Mining has not been available for nearly seventy five years. Included are details on the geology of this unique portion of North Eastern Oregon. This is the first part of a two book series on the area. Accompanying the text are rare photographs and historic maps. 8.5" X 11", 94 ppgs. Retail Price: $9.99

Field Identification of Minerals For Oregon Prospectors

Originally published in 1940, this important publication on Oregon Mining has not been available for nearly seventy five years. Included in this volume is an easy system for testing and identifying a wide range of minerals that might be found by prospectors, geologists and rockhounds in the State of Oregon, as well as in other locales. Topics include how to put together your own field testing kit and how to conduct rudimentary tests in the field. This volume is written in a clear and concise way to make it useful even for beginners. 8.5" X 11", 158 ppgs. Retail Price: $14.99

Idaho Mining Books

Gold in Idaho

Unavailable since the 1940's, this publication was originally compiled by the Idaho Bureau of Mines and includes details on gold mining in Idaho. Included is not only raw data on gold production in Idaho, but also valuable insight into where gold may be found in Idaho, as well as practical information on the gold bearing rocks and other geological features that will assist those looking for placer and lode gold in the State of Idaho. This volume also includes thirteen gold maps that greatly enhance the practical usability of the information contained in this small book detailing where to find gold in Idaho. 8.5" X 11", 72 ppgs. Retail Price: $9.99

Geology of the Couer D'Alene Mining District of Idaho

Unavailable since 1961, this publication was originally compiled by the Idaho Bureau of Mines and Geology and includes details on the mining of gold, silver and other minerals in the famous Coeur D'Alene Mining District in Northern Idaho. Included are details on the early history of the Coeur D'Alene Mining District, local tectonic settings, ore deposit features, information on the mineral belts of the Osburn Fault, as well as detailed information on the famous Bunker Hill Mine, the Dayrock Mine, Galena Mine, Lucky Friday Mine and the infamous Sunshine Mine. This volume also includes sixteen hard to find maps. 8.5" X 11", 70 ppgs. Retail Price: $9.99

The Gold Camps and Silver Cities of Idaho

Originally published in 1963, this important publication on Idaho Mining has not been available for nearly fifty years. Included are rare insights into the history of Idaho's Gold Rush, as well as the mad craze for silver in the Idaho Panhandle. Documented in fine detail are the early mining excitements at Boise Basin, at South Boise, in the Owyhees, at Deadwood, Long Valley, Stanley Basin and Robinson Bar, at Atlanta, on the famous Boise River, Volcano, Little Smokey, Banner, Boise Ridge, Hailey, Leesburg, Lemhi, Pearl, at South Mountain, Shoup and Ulysses, Yellow Jacket and Loon Creek. The story follows with the appearance of Chinese miners at the new mining camps on the Snake River, Black Pine, Yankee Fork, Bay Horse, Clayton, Heath, Seven Devils, Gibbonsville, Vienna and Sawtooth City. Also included are special sections on the Idaho Lead and Silver mines of the late 1800's, as well as the mining discoveries of the early 1900's that paved the way for Idaho's modern mining and mineral industry. Lavishly illustrated with rare historic photos, this volume provides a one of a kind documentary into Idaho's mining history that is sure to be enjoyed by not only modern miners and prospectors who still scour the hills in search of nature's treasures, but also those enjoy history and tromping through overgrown ghost towns and long abandoned mining camps. 8.5" X 11", 186 ppgs. Retail Price: $14.99

Utah Mining Books

Fluorite in Utah

Unavailable since 1954, this publication was originally compiled by the USGS, State of Utah and U.S. Atomic Energy Commission and details the mining of fluorspar, also known as fluorite in the State of Utah. Included are details on the geology and history of fluorspar (fluorite) mining in Utah, including details on where this unique gem mineral may be found in the State of Utah. 8.5" X 11", 60 ppgs. Retail Price: $8.99

California Mining Books

The Tertiary Gravels of the Sierra Nevada of California

Mining historian Kerby Jackson introduces us to a classic mining work by Waldemar Lindgren in this important re-issue of The Tertiary Gravels of the Sierra Nevada of California. Unavailable since 1911, this publication includes details on the gold bearing ancient river channels of the famous Sierra Nevada region of California. 8.5" X 11", 282 ppgs. Retail Price: $19.99

The Mother Lode Mining Region of California

Unavailable since 1900, this publication includes details on the gold mines of California's famous Mother Lode gold mining area. Included are details on the geology, history and important gold mines of the region, as well as insights into historic mining methods, mine timbering, mining machinery, mining bell signals and other details on how these mines operated. Also included are insights into the gold mines of the California Mother Lode that were in operation during the first sixty years of California's mining history. 8.5" X 11", 176 ppgs. Retail Price: $14.99

Lode Gold of the Klamath Mountains of Northern California and South West Oregon

Unavailable since 1971, this publication was originally compiled by Preston E. Hotz and includes details on the lode mining districts of Oregon and California's Klamath Mountains. Included are details on the geology, history and important lode mines of the French Gulch, Deadwood, Whiskeytown, Shasta, Redding, Muletown, South Fork, Old Diggings, Dog Creek (Delta), Bully Choop (Indian Creek), Harrison Gulch, Hayfork, Minersville, Trinity Center, Canyon Creek, East Fork, New River, Denny, Liberty (Black Bear), Cecilville, Callahan, Yreka, Fort Jones and Happy Camp mining districts in California, as well as the Ashland, Rogue River, Applegate, Illinois River, Takilma, Greenback, Galice, Silver Peak, Myrtle Creek and Mule Creek districts of South Western Oregon. Also included are insights into the mineralization and other characteristics of this important mining region. 8.5" X 11", 100 ppgs. Retail Price: $10.99

Mines and Mineral Resources of Shasta County, Siskiyou County, Trinity County, California

Unavailable since 1915, this publication was originally compiled by the California State Mining Bureau and includes details on the gold mines of this area of Northern California. Also included are insights into the mineralization and other characteristics of this important mining region, as well as the location of historic gold mines. 8.5" X 11", 204 ppgs. Retail Price: $19.99

Geology of the Yreka Quadrangle, Siskiyou County, California

Unavailable since 1977, this publication was originally compiled by Preston E. Hotz and includes details on the geology of the Yreka Quadrangle of Siskiyou County, California. Also included are insights into the mineralization and other characteristics of this important mining region. 8.5" X 11", 78 ppgs. Retail Price: $7.99

Mines of San Diego and Imperial Counties, California

Originally published in 1914, this important publication on California Mining has not been available for a century. This publication includes important information on the early gold mines of San Diego and Imperial County, which were some of the first gold fields mined in California by early Spanish and Mexican miners before the 49ers came on the scene. Included are not only details on early mining methods in the area, production statistics and geological information, but also the location of the early gold mines that helped make California "The Golden State". Also included are details on the mining of other minerals such as silver, lead, zinc, manganese, tungsten, vanadium, asbestos, barite, borax, cement, clay, dolomite, fluospar, gem stones, graphite, marble, salines, petroleum, stronium, talc and others. 8.5" X 11", 116 ppgs. Retail Price: $12.99

Mines of Sierra County, California

Unavailable since 1920, this publication was originally compiled by the California State Mining Bureau and includes details on the gold mines of Sierra County, California. Also included are insights into the mineralization and other characteristics of this important mining region, as well as the location of historic gold mines. 8.5" X 11", 156 ppgs. Retail Price: $19.99

Mines of Plumas County, California

Unavailable since 1918, this publication was originally compiled by the California State Mining Bureau and includes details on the gold mines of Plumas County, California. Also included are insights into the mineralization and other characteristics of this important mining region, as well as the location of historic gold mines. **8.5" X 11", 200 ppgs. Retail Price: $19.99**

Mines of El Dorado, Placer, Sacramento and Yuba Counties, California

Originally published in 1917, this important publication on California Mining has not been available for nearly a century. This publication includes important information on the early gold mines of El Dorado County, Placer County, Sacramento County and Yuba County, which were some of the first gold fields mined by the Forty-Niners during the California Gold Rush. Included are not only details on early mining methods in the area, production statistics and geological information, but also the location of the early gold mines that helped make California "The Golden State". Also included are insights into the early mining of chrome, copper and other minerals in this important mining area. **8.5" X 11", 204 ppgs. Retail Price: $19.99**

Mines of Los Angeles, Orange and Riverside Counties, California

Originally published in 1917, this important publication on California Mining has not been available for nearly a century. This publication includes important information on the early gold mines of Los Angeles County, Orange County and Riverside County, which were some of the first gold fields mined in California by early Spanish and Mexican miners before the 49ers came on the scene. Included are not only details on early mining methods in the area, production statistics and geological information, but also the location of the early gold mines that helped make California "The Golden State". **8.5" X 11", 146 ppgs. Retail Price: $12.99**

Mines of San Bernadino and Tulare Counties, California

Originally published in 1917, this important publication on California Mining has not been available for nearly a century. This publication includes important information on the early gold mines of San Bernadino and Tulare County, which were some of the first gold fields mined in California by early Spanish and Mexican miners before the 49ers came on the scene. Included are not only details on early mining methods in the area, production statistics and geological information, but also the location of the early gold mines that helped make California "The Golden State". Also included are details on the mining of other minerals such as copper, iron, lead, zinc, manganese, tungsten, vanadium, asbestos, barite, borax, cement, clay, dolomite, fluospar, gem stones, graphite, marble, salines, petroleum, stronium, talc and others. **8.5" X 11", 200 ppgs. Retail Price: $19.99**

Chromite Mining in The Klamath Mountains of California and Oregon

Unavailable since 1919, this publication was originally compiled by J.S. Diller of the United States Department of Geological Survey and includes details on the chromite mines of this area of Northern California and Southern Oregon. Also included are insights into the mineralization and other characteristics of this important mining region, as well as the location of historic mines. Also included are insights into chromite mining in Eastern Oregon and Montana. 8.5" X 11", 98 ppgs. Retail Price: $9.99

Mines and Mining in Amador, Calaveras and Tuolumne Counties, California

Unavailable since 1915, this publication was originally compiled by William Tucker and includes details on the mines and mineral resources of this important California mining area. Included are details on the geology, history and important gold mines of the region, as well as insights into other local mineral resources such as asbestos, clay, copper, talc, limestone and others. Also included are insights into the mineralization and other characteristics of this important portion of California's Mother Lode mining region. 8.5" X 11", 198 ppgs. Retail Price: $14.99

The Cerro Gordo Mining District of Inyo County California

Unavailable since 1963, this publication was originally compiled by the United States Department of Interior. Included are insights into the mineralization and other characteristics of this important mining region of Southern California. Topics include the mining of gold and silver in this important mining district in Inyo County, California, including details on the history, production and locations of the Cerro Gordo Mine, the Morning Star Mine, Estelle Tunnel, Charles Lease Tunnel, Ignacio, Hart, Crosscut Tunnel, Sunset, Upper Newtown, Newtown, Ella, Perseverance, Newsboy, Belmont and other silver and gold mines in the Cerro Gordo Mining District. This volume also includes important insights into the fossil record, geologic formations, faults and other aspects of economic geology in this California mining district. 8.5" X 11", 104 ppgs. Retail Price: $10.99

Mining in Butte, Lassen, Modoc, Sutter and Tehama Counties of California

Unavailable since 1917, this publication was originally compiled by the United States Department of Interior. Included are insights into the mineralization and other characteristics of this important mining region of California. Topics include the mining of asbestos, chromite, gold, diamonds and manganese in Butte County, the mining of gold and copper in the Hayden Hill and Diamond Mountain mining districts of Lassen County, the mining of coal, salt, copper and gold in the High Grade and Winters mining districts of Modoc County, gold mining in Sutter County and the mining of gold, chromite, manganese and copper in Tehama County. This volume also includes the production records and locations of numerous mines in this important mining region. 8.5" X 11", 114 ppgs. Retail Price: $11.99

Alaska Mining Books

Ore Deposits of the Willow Creek Mining District, Alaska

Unavailable since 1954, this hard to find publication includes valuable insights into the Willow Creek Mining District near Hatcher Pass in Alaska. The publication includes insights into the history, geology and locations of the well known mines in the area, including the Gold Cord, Independence, Fern, Mabel, Lonesome, Snowbird, Schroff-O'Neil, High Grade, Marion Twin, Thorpe, Webfoot, Kelly-Willow, Lane, Holland and others. 8.5" X 11", 96 ppgs. Retail Price: $9.99

Arizona Mining Books

Mines and Mining in Northern Yuma County Arizona

Originally published in 1911, this important publication on Arizona Mining has not been available for over a hundred years. Included are rare insights into the gold, silver, copper and quicksilver mines of Yuma County, Arizona together with hard to find maps and photographs. Some of the mines and mining districts featured include the Planet Copper Mine, Mineral Hill, the Clara Consolidated Mine, Viati Mine, Copper Basin prospect, Bowman Mine, Quartz King, Billy Mack, Carnation, the Wardwell and Osbourne, Valensuella Copper, the Mariquita, Colonial Mine, the French American, the New York-Plomosa, Guadalupe, Lead Camp, Mudersbach Copper Camp, Yellow Bird, the Arizona Northern (Salome Strike), Bonanza (Harqua Hala), Golden Eagle, Hercules, Socorro and others. 8.5" X 11", 144 ppgs. Retail Price: $11.99

The Aravaipa and Stanley Mining Districts of Graham County Arizona

Originally published in 1925, this important publication on Arizona Mining has not been available for nearly ninety years. Included are rare insights into the gold and silver mines of these two important mining districts, together with hard to find maps. 8.5" X 11", 140 ppgs. Retail Price: $11.99

Gold in the Gold Basin and Lost Basin Mining Districts of Mohave County, Arizona

This volume contains rare insights into the geology and gold mineralization of the Gold Basin and Lost Basin Mining Districts of Mohave County, Arizona that will be of benefit to miners and prospectors. Also included is a significant body of information on the gold mines and prospects of this portion of Arizona. This volume is lavishly illustrated with rare photos and mining maps. 8.5" X 11", 188 ppgs. Retail Price: $19.99

Mines of the Jerome and Bradshaw Mountains of Arizona

This important publication on Arizona Mining has not been available for ninety years. This volume contains rare insights into the geology and ore deposits of the Jerome and Bradshaw Mountains of Arizona that will be of benefit to miners and prospectors who work those areas. Included is a significant body of information on the mines and prospects of the Verde, Black Hills, Cherry Creek, Prescott, Walker, Groom Creek, Hassayampa, Bigbug, Turkey Creek, Agua Fria, Black Canyon, Peck, Tiger, Pine Grove, Bradshaw, Tintop, Humbug and Castle Creek Mining Districts. This volume is lavishly illustrated with rare photos and mining maps. 8.5" X 11", 218 ppgs. Retail Price: $19.99

The Ajo Mining District of Pima County Arizona

This important publication on Arizona Mining has not been available for nearly seventy years. This volume contains rare insights into the geology and mineralization of the Ajo Mining District in Pima County, Arizona and in particular the famous New Cornelia Mine.   8.5" X 11", 126 ppgs. Retail Price: $11.99

Mining in the Santa Rita and Patagonia Mountains of Arizona

Originally published in 1915, this important publication on Arizona Mining has not been available for nearly a century. Included are rare insights into hundreds of gold, silver, copper and other mines in this famous Arizona mining area. Details include the locations, geology, history, production and other facts of the mines of this region. 8.5" X 11", 394 ppgs. Retail Price: $24.99

Montana Mining Books

A History of Butte Montana: The World's Greatest Mining Camp

First published in 1900 by H.C. Freeman, this important publication sheds a bright light on one of the most important mining areas in the history of The West. Together with his insights, as well as rare photographs of the periods, Harry Freeman describes Butte and its vicinity from its early beginnings, right up to its flush years when copper flowed from its mines like a river. At the time of publication, Butte, Montana was known worldwide as "The Richest Mining Spot On Earth" and produced not only vast amounts of copper, but also silver, gold and other metals from its mines. Freeman illustrates, with great detail, the most important mines in the vicinity of Butte, providing rare details on their owners, their history and most importantly, how the mines operated and how their treasures were extracted. Of particular interest are the dozens of rare photographs that depict mines such as the famous Anaconda, the Silver Bow, the Smoke House, Moose, Paulin, Buffalo, Little Minah, the Mountain Consolidated, West Greyrock, Cora, the Green Mountain, Diamond, Bell, Parnell, the Neversweat, Nipper, Original and many others.   8.5" X 11", 142 ppgs. Retail Price: $12.99

The Butte Mining District of Montana

This important publication on Montana Mining has not been available for over a century. Included are rare insights into the gold, copper and silver mines of Butte, Montana together with hard to find maps and photographs. Some of the topics include the early history of gold, silver and copper mining in the Butte area, insight into the geology of its mining areas, the local distribution of gold, silver and copper ores, as well their composition and how to identify them. Also included are detailed facts about the mines in the Butte Mining District, including the famous Anaconda Mine, Gagnon, Parrot, Blue Vein, Moscow, Poulin, Stella, Buffalo, Green Mountain, Wake Up Jim, the Diamond-Bell Group, Mountain Consolidated, East Greyrock, West Greyrock, Snowball, Corra, Speculator, Adirondack, Miners Union, the Jessie-Edith May Group, Otisco, Iduna, Colorado, Lizzie, Cambers, Anderson, Hesperus, Preferencia and dozens of others. 8.5" X 11", 298 ppgs. Retail Price: $24.99

Mines of the Helena Mining Region of Montana

This important publication on Montana Mining has not been available for over a century. Included are rare insights into the gold, copper and silver mines of the vicinity of Helena, Montana, including the Marysville Mining District, Elliston Mining District, Rimini Mining District, Helena Mining District, Clancy Mining District, Wickes Mining District, Boulder and Basin Mining Districts and the Elkhorn Mining District. Some of the topics include the early history of gold, silver and copper mining in the Helena area, insight into the geology of its mining areas, the local distribution of gold, silver and copper ores, as well their composition and how to identify them. Also included are detailed facts, history, geology and locations of over one hundred gold, silver and copper mines in the area . 8.5" X 11", 162 ppgs, Retail Price: $14.99

Mines and Geology of the Garnet Range of Montana

This important publication on Montana Mining has not been available for over a century. Included are rare insights into the gold, copper and silver mines of the vicinity of this important mining area of Montana. Some of the topics include the early history of gold, silver and copper mining in the Garnet Mountains, insight into the geology of its mining areas, the local distribution of gold, silver and copper ores, as well their composition and how to identify them. Also included are detailed facts, history, geology and locations of numerous gold, silver and copper mines in the area . 8.5" X 11", 100 ppgs, Retail Price: $11.99

Mines and Geology of the Philipsburg Quadrangle of Montana

This important publication on Montana Mining has not been available for over a century. Included are rare insights into the gold, copper and silver mines of the vicinity of this important mining area of Montana. Some of the topics include the early history of gold, silver and copper mining in the Philipsburg Quadrangle, insight into the geology of its mining areas, the local distribution of gold, silver and copper ores, as well their composition and how to identify them. Also included are detailed facts, history, geology and locations of over one hundred gold, silver and copper mines in the area 8.5" X 11", 290 ppgs, Retail Price: $24.99

Geology of the Marysville Mining District of Montana

Included are rare insights into the mining geology of the Marysville Mining District. Some of the topics include the early history of gold, silver and copper mining in the area, insight into the geology of its mining areas, the local distribution of gold, silver and copper ores, as well their composition and how to identify them. Also included are detailed facts, history, geology and locations of gold, silver and copper mines in the area 8.5" X 11", 198 ppgs, Retail Price: $19.99

More Mining Books

Prospecting and Developing A Small Mine

Topics covered include the classification of varying ores, how to take a proper ore sample, the proper reduction of ore samples, alluvial sampling, how to understand geology as it is applied to prospecting and mining, prospecting procedures, methods of ore treatment, the application of drilling and blasting in a small mine and other topics that the small scale miner will find of benefit. 8.5" X 11", 112 ppgs, Retail Price: $11.99

Timbering For Small Underground Mines

Topics covered include the selection of caps and posts, the treatment of mine timbers, how to install mine timbers, repairing damaged timbers, use of drift supports, headboards, squeeze sets, ore chute construction, mine cribbing, square set timbering methods, the use of steel and concrete sets and other topics that the small underground miner will find of benefit. This volume also includes twenty eight illustrations depicting the proper construction of mine timbering and support systems that greatly enhance the practical usability of the information contained in this small book. 8.5" X 11", 88 ppgs. Retail Price: $10.99

Timbering and Mining

A classic mining publication on Hard Rock Mining by W.H. Storms. Unavailable since 1909, this rare publication provides an in depth look at American methods of underground mine timbering and mining methods. Topics include the selection and preservation of mine timbers, drifting and drift sets, driving in running ground, structural steel in mine workings, timbering drifts in gravel mines, timbering methods for driving shafts, positioning drill holes in shafts, timbering stations at shafts, drainage, mining large ore bodies by means of open cuts or by the "Glory Hole" system, stoping out ore in flat or low lying veins, use of the "Caving System", stoping in swelling ground, how to stope out large ore bodies, Square Set timbering on the Comstock and its modifications by California miners, the construction of ore chutes, stoping ore bodies by use of the "Block System", how to work dangerous ground, information on the "Delprat System" of stoping without mine timbers, construction and use of headframes and much more. This volume provides a reference into not only practical methods of mining and timbering that may be employed in narrow vein mining by small miners today, but also rare insights into how mines were being worked at the turn of the 19th Century. 8.5" X 11", 288 ppgs. Retail Price: $24.99

A Study of Ore Deposits For The Practical Miner

Mining historian Kerby Jackson introduces us to a classic mining publication on ore deposits by J.P. Wallace. First published in 1908, it has been unavailable for over a century. Included are important insights into the properties of minerals and their identification, on the occurrence and origin of gold, on gold alloys, insights into gold bearing sulfides such as pyrites and arsenopyrites, on gold bearing vanadium, gold and silver tellurides, lead and mercury tellurides, on silver ores, platinum and iridium, mercury ores, copper ores, lead ores, zinc ores, iron ores, chromium ores, manganese ores, nickel ores, tin ores, tungsten ores and others. Also included are facts regarding rock forming minerals, their composition and occurrences, on igneous, sedimentary, metamorphic and intrusive rocks, as well as how they are geologically disturbed by dikes, flows and faults, as well as the effects of these geologic actions and why they are important to the miner. Written specifically with the common miner and prospector in mind, the book will help to unlock the earth's hidden wealth for you and is written in a simple and concise language that anyone can understand. **8.5" X 11", 366 ppgs. Retail Price, $24.99**

Mine Drainage

Unavailable since 1896, this rare publication provides an in depth look at American methods of underground mine drainage and mining pump systems. This volume provides a reference into not only practical methods of mining drainage that may be employed in narrow vein mining by small miners today, but also rare insights into how mines were being worked at the turn of the 19th Century. **8.5" X 11", 218 ppgs. Retail Price, $24.99**

Fire Assaying Gold, Silver and Lead Ores

Unavailable since 1907, this important publication was originally published by the Mining and Scientific Press and was designed to introduce miners and prospectors of gold, silver and lead to the art of fire assaying. Topics include the fire assaying of ores and products containing gold, silver and lead; the sampling and preparation of ore for an assay; care of the assay office, assay furnaces; crucibles and scorifiers; assay balances; metallic ores; scorification assays; cupelling; parting' crucible assays, the roasting of ores and more. This classic provides a time honored method of assaying put forward in a clear, concise and easy to understand language that will make it a benefit to even beginners. **8.5" X 11", 96 ppgs. Retail Price, $11.99**

www.ingramcontent.com/pod-product-compliance
Lightning Source LLC
Chambersburg PA
CBHW081831170526
45167CB00007B/2792